L'Arte Vetraria

The Art of Glass

Volume III of III

L'Arte Vetraria
The Art of Glass
by Antonio Neri
Volume III of III

Translated & Annotated by
Paul Engle

Heiden & Engle

Editorial, Sales, & Customer Service Office:

Heiden & Engle
P.O. Box 451
Hubbardston, MA 01452

Copyright © 2003-2007 by Heiden & Engle

All rights reserved. No part of the material protected by this copyright notice may be reproduced or utilized in any form, electronic or mechanical, including photocopying, recording, or by any information storage and retrieval system, without written permission of the copyright owner.

Publication Data

Engle, Paul, 1959-
    L'Arte Vetraria : The Art of Glass / Paul Engle. - 1st ed.
    Includes bibliographical references and index.
    Volume III of a 3 part set:
    ISBN 0-9743529-3-4 (Volume III, pbk)
    ISBN 0-9743529-0-X (3 Volume Set)

for Lori

86	*An Oriental Sapphire Loaded with Color.*	11
87	*Oriental Garnet. (I)*	12
88	*A Deeper Oriental Garnet. (II)*	12
89	*Another Beautiful Garnet. (III)*	13
90	*Advice about Pastes, and their Colors.*	14
91	*A Marvelous Way, No Longer Used, To Make the Above Paste, and to Imitate Every Type of Gem.*	15
92	*A Way to Make Hard Pastes in All the Colors.*	18

Book Six

In which I show the way to make all of the goldsmiths' enamels to fire over gold in various colors. Included are the rules, the colorant materials, and the methods to make the fires for such enamels with exquisite diligence. I present the demonstrations as clearly as possible for this subject. — 21

93	*The Material to Make All Enamels.*	22
94	*Milk White Enamel [Lattimo].*	23
95	*[Two] Turquoise Enamels.*	24
96	*Another Blue Enamel.*	26
97	*Green Enamel. (I)*	26
98	*Another Green Enamel. (II)*	27
99	*Another Green Enamel. (III)*	27
100	*Black Enamel. (I)*	28

CONTENTS

Acknowledgements   xiii
Forward   xv

THE ART OF GLASS

Book Five

75   *In which is shown the true way to make pastes for Emerald, Topaz, Chrysolite, Jacinth, Garnet, Sapphire, Aquamarine, and other colors, of so much grace, and beauty, that they will surpass the natural stones in everything except hardness. With a new chemical method never before used, to make the above pastes of Isaac Hollandus, which by far surpass in beauty, and color of all other pastes made until today.*   1

76   *The Way to Prepare Rock Crystal.*   2

77   *The Way to Make Oriental Emerald. (I)*   4

78   *An Emerald More Charged with Color. (II)*   6

79   *To Make a More Graceful Emerald Paste. (III)*   7

80   *Another Most Attractive Emerald. (IV)*   8

81   *Oriental Topaz.*   9

82   *Oriental Chrysolite.*   9

83   *A Celestial Color.*   10

84   *A Celestial Color with Violet.*   10

85   *Oriental Sapphire.*   11

101	*Another Black Enamel. (II)*	29
102	*Another Black Enamel. (III)*	29
103	*Red Wine Colored Enamel.*	30
104	*Purple Enamel.*	31
105	*Yellow Enamel.*	31
106	*A Celestial Blue Enamel.*	32
107	*A Violet Enamel.*	33

Book Seven

In which I show the way to extract a yellow lake for painters from broom flowers, and all the other colors. With another way to extract lakes of red, green, blue, purple, and of all the colors from all types of plants and flowers. ...Also shown is the way to make transparent red glass, and other things that today seem obsolete and disused, like the way to make rosichiero for enameling over gold. 35

108	*A Yellow Lake from Broom Flowers, for Paint.*	36
109	*A Way to Extract Lake from Poppies, Blue Irises, Red Roses, Violet Roses, and From All Kinds of Green Plants.*	37
110	*A Way to Extract the Lake and Color for Painting, from Orange Blossoms, Red Poppies, Blue Irises, Ordinary Violets, Red Violets, Carnations, Red Roses, Borage Flowers, Day Lilies, Irises, and From Flowers of Any Desired Color, and the Greens of the Mallow, the Pimpernel, and All the Plants.*	38
111	*An Azure [Paint] like Alemagna Blue.*	39

112	The Way to Color Faded Natural Turquoise.	40
113	A Mixture to Make [Mirror] Spheres.	40
114	The Way to Tint Glass Balls, and Others Vessels of Clear Glass, From the Inside, In All Kinds of Colors, So They Will Imitate Natural Stones.	41
115	Ultramarine Blue.	42
116	[Prepared Wool Shearings to Make a] Lake of Kermes for Painters.	45
117	An Elixir for extracting the color from Kermes.	46
118	Very Beautiful Lakes of Brazil Wood, and Madder Root.	49
119	Lake of Kermes another Easier Way.	50
120	Transparent Red in Glass.	51
121	Red Like Blood.	52
122	Balas Color.	52
123	To extract the Spirit of Saturn, Which Serves Many Uses in Enamels and Glasses.	53
124	Rosichiero to Enamel Gold.	54
125	Rosichiero for Gold by Another Method.	54
126	How to Fix Sulfur for the Above Described Work.	56
127	A Glass as Red as Blood, Which Can Serve as Rosichiero.	56
128	A Proven Way to Make Rosichiero.	57
129	Transparent Red.	58
130	The Way to Fix Sulfur for Rosichiero to Enamel Gold.	59

131	*Vitriol of Venus, Continued from the End of Chapter 31.*	59
132	*Vitriolo of Copper, Also Known As Vitriol of Venus, Made Without Corrosives, From Which is Taken the Truly Vivid Light Blue, a Marvelous Thing.*	61
133	*The Way to Extract Vitriol From the Above Colored Water.*	64

Original Table of Contents — 67

Permission to Publish — 77

Glossary — 79

Biography — 107

Bibliography — 111

Picture Credits — 112

Appendix A *An 1870 Letter by G. F. Rodwell* — 113

Appendix B *Weights, Measures, and Miscellany* — 121

Index — 123

"The curious and noble spirit will be able to get a marvelous red from gold"

-Antonio Neri, chap. 90

ACKNOWLEDGEMENTS

First and foremost I would like to recognize the custodians of the written history of glass; the librarians. It is not an exaggeration to say that this project would have never gotten off the ground without the kind attention, patience, and helpful suggestions of these largely unsung heroes in institutions around the world. Particularly I would like to thank Gail Bardhan at the Corning Rakow Museum Library, and Niki Pollock at the Glasgow University Library. Also many thanks to the staffs of the rare books departments at The University of Massachusetts at Amherst, Cornell University, Yale University, the University of Texas, and the United States Library of Congress. While important assistance has been lent from web sites too numerous to mention, I would like to single out the development and support teams for the Bibliothèque Nationale de France, Biblioteca Nazionale di Firenze, Istituto e Museo di Storia della Scienza, The Medici Archive Project, the Galileo Project, the Vatican Library, the Tufts University Persius Latin site, the Alchemy Web Site, Garzanti Linguistica, the Catholic Encyclopedia site and of course Google. Many thanks also to Jon Eklund at the Smithsonian Institution, and especially to Elizabeth and Julia Whitehouse at Whitehouse books, for unwittingly setting the wheels in motion by selling a rare 1661 second edition of *L'Arte Vetraria* to my wife Lori and me several years ago. Similarly, many fine authors and editors have helped to shape my understanding and appreciation of the late Renaissance, chief among them, the late Dean of Italian glass historians, Luigi Zecchin. Thanks also to Patrick McCray, David Freedberg, Philip Ball, and Jacob Burkhardt. For their invaluable efforts to bring the scarce first edition to light through contemporary reprints, I thank Rosa Barovier Mentasti, and also Ferdinando Abbri with Giunti, Neri's original printer, still going, four hundred years later; a firm that was already in the book business when Cristopher Columbus set sail for the new world... simply amazing, or as the Reverend Priest himself might have quipped: 'Portano la palma'. I would also like to thank John Woodhouse at the University of Oxford, and Kenneth Brown at the University of Calgary for their helpful clarifications, and Emilio Santini, a prince among glass artists, for his discussion and encouragement of the project. Finally, an especially warm thank you goes to my partner Lori Engle, who, by infecting me with her passion for glass and sharing this adventure, has helped make the impossible a reality.

Volume II

In addition to the above mentioned, many of whom have continued their kind support for the second volume, I would like to acknowledge Deborah K. Zinn of the Beads-L Online Knowledgebase, Deborah Barlow Smedstad at the Museum of Fine Arts, Boston, Carol Kaufmann of the Iziko Museum of Cape Town South Africa, Sonny Maley at the University of Glasgow Library, and Margret Carey of the Bead Study Trust in London. They have all contributed in some way to my continuing education on the history of glass in the sixteenth and seventeenth century, and to tying up the myriad loose ends in this project. A special thanks goes to Scottish author Adam McLean, for the Alchemy website, and his prolific writing on Hermetic tradition. This is a very complex subject, and his excellent work has been a beacon of light in navigating the two Neri manuscripts at Glasgow.

Volume III

Thanks to so many of the above for their continued assistance especially to my wife Lori Engle. Also to Dedo von Kerssenbrock-Krosigk, at the Corning Museum, Dietmar Kühlmorgen, Phil Joinville at DPR, and to my tireless supporter Louise Erskine, a real angel.

FOREWORD

This third and final volume covers chapters 75 through 133 of *L'Arte Vetraria*. Book 5 details the creation of imitation gems, that "... by far surpass the beauty and color of all other pastes made until today". Book 6 covers enamels, both clear and opaque, with advice for varying the intensity of the colors, and Book 7 contains a variety of recipes, some for glass, some for enamels, and others that range from paints, to restoring faded turquoise, to casting white bronze mirrors.

In this volume the glossary has been expanded again to include the specialized terms used in this part of Neri's work, and weighs in at 556 entries. Appendix B contains useful weights and measures information, and the Biography outlines short descriptions of the major characters surrounding Neri's life and the book. Appendix A is an annotated reprint of a letter which appeared in the Sept. 8, 1870 issue of *Nature*, written by George Rodwell, discussing the emergence of an unpublished manuscript by Antonio Neri, which now resides at Glasgow University (*MS Ferguson 168*).

"Perform this operation under a large chimney, and when the fumes begin to assault you, it is best to leave the room. This smoke is most injurious, and could be deadly; therefore, you should see that no one inhales it in any way, because it would do very great damage. When all fumes pass, you should nevertheless leave the crucibles in a low fire, until it goes out completely."

- Antonio Neri, chap.73

WARNING

The recipes in *L'Arte Vetraria* contain extremely hazardous methods and materials. None of these procedures should ever be attempted under any circumstances outside of laboratory conditions. One particular danger (but certainly not the only one) is Neri's frequent use of mercury in the glass melt. Liver, kidney, heart, lung and irreversible brain damage are the result of exposure to even small quantities of mercury that has been vaporized. In the sixteenth and seventeenth centuries, arsenic, lead, mercury, and other virulent poisons were handled in ways that are considered reckless by current standards.

A safe and responsible investigation of the recipes in *L'Arte Vetraria* would at minimum involve a world-class glassmaking laboratory facility, a staff of well trained knowledgeable professionals, verifiable monitoring equipment, and a ventilation system with fume scrubbers to protect both workers and the environment.

Please do not endanger yourself or those around you by attempting any of these recipes. The lack of proper equipment and precautions could easily result in poisoning, permanent injury, or death. Please, please, please, if you are a novice, and you hear the glass-muse calling, contact a reputable organization that can give you appropriate advice on how to get started. GAS - the Glass Art Society (www.glassart.org), and ISGB - the International Society of Glass Beadmakers (www.isgb.org) are two.

"Within this lies all the expertise; it is the principal skill in the mastery of this operation. When you grind the crystal optimally, you can make artificial gems that in beauty, color, clarity, splendor, and polish by far exceed the natural, true ones."

- Antonio Neri, chap. 76

The Fifth Book of The Art of Glass

By PRIEST ANTONIO NERI

In which I show the true way to make pastes for emerald, topaz, chrysolite, jacinth, garnet, sapphire, aquamarine, and other colors, of so much grace and beauty, that they will surpass the natural stones in everything except hardness.

Included is a new chemical method never before used, to make the above pastes [by the method] of Isaac Hollandus, which by far surpass the beauty and color of all other pastes made until today.

~ 75 ~

I believe the knowledge to imitate emerald, topaz, chrysolite, sapphire, garnet, and nearly every kind of gem is something that few would not desire and seek with all earnestness. It is quite appealing and compelling to imitate their color, splendor, grace, and polish (save hardness) with a level of perfection that exceeds the natural oriental stones.

Therefore, in the present book I describe the true way to make them, with all the conditions and instructions necessary, since not only will they be similar, but they will surpass all the qualities of the natural stones except hardness. There is no doubt that anyone who does the work with diligence will accomplish much more than simply what I describe.

Above all is this wonderful invention. A new way practiced by me, with the doctrine taken from Isaac Hollandus, in which paste jewels of so much grace, beauty, and perfection are made, that they seem nearly impossible to describe, and hard to believe.

It is quite true that the work is somewhat long and laborious. Nevertheless, I who have done it many times, say it is altogether easy and plain to do. After everything, upon completion, all the labor and expense employed in this work will seem small and insubstantial.

~ 76 ~
The Way to Prepare Rock Crystal

You should have rock crystal of the greatest possible clarity and beauty. Free it of any pieces of flint, chalcedony, quartzite, or other hard stones that may vitrify well but do not produce the clear lucidity and shine, or take the marvelous polish that rock crystal does in this procedure. These [other] stones always have an essence of the earth, and obscurity in them, while the crystal always has an essence of the air, and transparency. It closely approaches the quality and natural character of [true] gems, especially when oriental rock crystal is used; it shows far more beautiful effects than either Italian or German crystal.

Therefore, you should dress the crystal clean, and put it into a covered crucible. Keep it among the burning coals, where it will be inflamed thoroughly and fired well. At this point, quickly throw the crystal into a sizeable pan full of fresh clear water.

Once it cools, remove it from the water, and return it to the crucible to calcine and inflame in the coals. Always take care to cover the crucible because no embers or ash should get inside; this procedure requires [the material] to be made with great cleanliness, and with exquisite care. Again, when it is inflamed well, throw the glass into the pan of water, which should be changed and always fresh and clean. Repeat this calcination at least twelve times. Now the crystal is well calcined, and ready to grind on the porphyry to reduce it to the finest powder; like sifted grain flour. Take the crystal, calcined and dried, and grind it as usual over a porphyry stone with a small muller of porphyry.

It will crumble, and decompose like refined sugar. Be sure never to crush it in a bronze mortar, since it will not be possible to get the desired outcome. In such a case, it will take the color of the copper [from the mortar] and the iron from the pestle and you will produce nothing other than emerald green. Take great care in crushing and grinding it on the porphyry stone. It is very important here to use great diligence and patience. Grind it one time, grind it again, and then make another pass. Always put only a little onto the porphyry at once, which is to say half a spoonful at a time. Grind most of it impalpably, then in turn regrind it, and finally take a third pass and always on a porphyry stone as described above.

It should be ground until when it is touched with the fingers it does not have any feel of grittiness, but it is, in all and for all, like finely sifted grain flour. Any grit will give the paste an ugly appearance, and will not closely resemble natural gems. Indeed the work will be ugly, and imperfect, so use diligence, and great patience. In this way, you will grind the crystal optimally. Within this lies all the expertise; it is the principal skill in the mastery of this operation. When you grind the crystal optimally, you can make artificial gems that in beauty, color, clarity, splendor, and polish by far exceed the natural, true ones. So again, I repeat; you must grind the crystal perfectly, and of this have a good amount, enough to be able to make of all the colors, because this is the basic material to make all the artificial gems.

~ 77 ~
The Way to Make Oriental Emerald

Take 2 oz of rock crystal ground impalpably, as I have described above [chapter 76]. Add 4 oz of ordinary minium, blend these powders together thoroughly, and unite them well so they incorporate. In all you will have 6 oz of material, to which add 8 grains per ounce of good verdigris, finely ground. Therefore, in all give 2 pwt [pennyweight] of ground verdigris to the material. Stir in 8 grains of crocus martis made with vinegar, as described in the seventeenth chapter. Mix this thoroughly, and then blend diligently with the aforementioned materials.

Put them in a good crucible that is resistant to fire, and is large enough for the entire mass of powder. Fill it, leaving a space the depth of one finger. Cover this crucible over with a terracotta lid, and lute it securely, then let it dry. Now put it in a pottery furnace to bake along with the pots, plates, and other vessels. Place it in an area of the furnace that will develop a good fire. In this heat and flame, the material purifies and cooks thoroughly. Leave it to stand in as strong a fire as bakes the pottery vessels, and then remove it from the furnace.

Open the crucible, and expose beautifully vitrified material of a lovely emerald color. In grand works and in the form of jewels it will seem superior to oriental emeralds from ancient rock. If the first time this paste is not sufficiently cooked and purified for lack of heat, re-fire it another time to purify. Find the sign of purification by looking in from the top, take care not to break the crucible but only lift the cover. If the paste is purified and transparent through to the bottom then it is good. If it does not sufficiently clarify, and it is opaque, then re-lute the cover to the crucible, and return it [to the furnace] another time to purify.

Take care never to break up the crucible before you look at

the paste, and make sure that it is cooked and purified thoroughly. If you break the crucible when the paste is not cooked, and you recover it into another crucible, then it will clean up, however you will see it filled with specs, and it will make for ugly and imperfect work. This is important to avoid.

If you do not have access to a pottery furnace, you can use a purpose-built kiln. Fire it for 24 hours, which will make the work perfect. Firing many crucibles of various colors at the same time makes for a good savings of work. Build the fire with hardwood, which is to say well-dried oak; green wood will not work well. The fire should be continuous and never allowed to slacken, because then the work will be imperfect.

In Antwerp, I built a furnace that held twenty glass-pots of various colors, and when fired for 24 hours everything fused and purified. For large loads, continue the fire for 6 hours more, with the assurance that the materials will cook thoroughly. This way, you will not consume excessive firewood. Be sure that the crucibles are of good quality and tempered to the fire.

Mount these pastes like the ordinary gems, and work them the same way in all respects. They will take on the polish and luster of the natural ones, and you can mount them in gold, backed with foil as usual. With these doses, the pastes will be harder than ordinary.

~ 78 ~
An Emerald More Charged with Color

Take 1 oz of rock crystal, ground impalpably as described above [chapter 76]. Add 6½ oz of ordinary minium, and as before blend them together well. To this material you will give verdigris in the proportion of 10 grains per ounce, along with [a fixed] 10 grains of crocus martis made with vinegar [chapter 17], totaling an additional 3 pwt and 13 grains.

Blend this material together thoroughly as was described for the other green [chapter 77], in all and for all. As usual, put it in a crucible that will withstand the fire. Leave a space of one finger width under the lid, lute it as usual, and cook the paste as described for the other [green].

This will be a color well charged with marvelous oriental emerald. In small works set in gold with foil, it will be beautiful beyond measure. Cook this paste more than the other one, in order to consume all the imperfections, which are usually in the lead. This paste with these doses will be softer than the above recipe, but the color will be most beautiful.

~ 79 ~
To Make a More Graceful Emerald Paste

Take 2 oz of [rock] crystal ground as before, and 7 oz of ordinary minium. Mix them together well, as usual like the others. Add to them a proportion of 10 grains of verdigris per ounce. Therefore, to these 9 oz of material, there will be in all, 3 pwt and 18 grains of well ground verdigris. Also, add a a single dose of 10 grains of crocus martis made with vinegar [chapter 17].

Blend these together with the minium and crystal, and put it into a crucible that is resistant to the fire. Lute it as before like the others, and then cook it as usual. Use it in small works because it will be loaded with color.

Making it this way you will have a graceful emerald, very sightly, but not very hard due to the great amount of lead. It is necessary, therefore, to let it cook in the fire more than usual, so that the fire consumes the lividness and unctuosity that by nature the lead confers, this produces a most sightly emerald.

~ 80 ~

Another Most Attractive Emerald

Take 2 oz of rock crystal ground impalpably, as above [chapter 76]. Add 6 oz of ordinary minium, for a total of 8 oz in all. Mix these powders together thoroughly, and to each ounce of material add 10 grains of well-ground and pulverized verdigris. In all, you will add 3 pwt and 8 grains of well ground verdigris to this material.

Blend it well with the above powder, and then put it in a crucible leaving a space the width of a finger. Cover it with a terracotta lid and lute it. Cook it in a furnace or kiln, as has already been described. Facet this material, and incorporate it into your work. This will be a most attractive oriental emerald color.

~ 81 ~
Oriental Topaz

Take 2 oz of rock crystal ground impalpably as above [chapter 76], and add 7 oz of ordinary minium. Mix them together thoroughly, and put them into a crucible that will resist fire. Leave a space the width of a finger, which is to say the crucible must remain empty one finger's width at the top.

These pastes always froth and swell, and will cling to the cover, and make an ugly sight. Therefore, be diligent in this, because when you top them off, they bubble and swell, spewing out of the crucible, making the paste imperfect. So take the precaution of leaving space, and then lute the crucible as usual. Cook it as was described in the other [recipes] in a pottery furnace, or else in a purpose-built kiln. A wonderful oriental topaz color will result, with which you can make any type of work that you might want.

~ 82 ~
Oriental Chrysolite

Take 2 oz of ground rock crystal as before, and 8 oz of ordinary minium. In all, there will be 10 oz to blend and mix thoroughly. Give them 12 grains of crocus martis made with vinegar [chapter 17], and then put them into a crucible that is resistant to fire, and leave a space of one finger as before.

Lute it as usual, cook it in a furnace as usual, and put other pastes in along with it. The fire consumes all imperfections of the lead resulting in an oriental chrysolite color, of all beauty, that in works [backed] with foil will produce marvelous effects.

~ 83 ~
A Celestial Color

Take 2 oz of rock crystal, ground impalpably as before, and add 5 oz of ordinary minium. In all, there will be 7 oz of material. Blend it together thoroughly, and to this add a proportion of 3 grains per ounce of prepared and ground zaffer. Therefore, in all give 21 grains of zaffer to the 7 oz, and blend them thoroughly, as was said for other pastes above. Put it into a crucible, with the aforementioned observations. Consign it to a pottery furnace, or purpose-built kiln. You will have a most graceful celestial color.

~ 84 ~
A Celestial Color with Violet

Take 2 oz of rock crystal, ground impalpably as usual, and 4½ oz of ordinary minium. Blend everything together for a total of 6½ oz, and to this material give a proportion of 4 grains of blue enamel that painters use. In all, there will be 1 pwt and 2 grains of enamel. Blend it well with the other powders and put it into a crucible. Lute it, and cook it in a potter's furnace, or even better in a furnace built for the purpose. Works [made] with this paste will have a most beautiful violet tint in a very graceful celestial color.

~ 85 ~
Oriental Sapphire

Take 2 oz of rock crystal, ground to an impalpable state, as before. Add 6 oz of ordinary minium, for a total of 8 oz in all. Blend and mix them thoroughly, and to every ounce of this material give a proportion of 5 grains of prepared zaffer, as before, so that in all there will be 1 pwt and 16 grains of zaffer. Into the zaffer, mix 6 grains of prepared Piedmont manganese.

Blend them all together thoroughly, mix them with the above powder, and put them into a crucible as described for the other pastes. Cook it and the result will be an oriental sapphire that has a most attractive violet tint.

~ 86 ~
An Oriental Sapphire Loaded with Color

Take 2 oz of rock crystal ground impalpably, as was described in its place [chapter 76]. Add 5 oz of ordinary minium, which is 7 oz in all. To each ounce of this material, give a proportion of 6 grains of prepared zaffer, which will be 1 pwt and 18 grains [of zaffer] in all. Into this zaffer, blend 8 grains of prepared Piedmont manganese. Unite these things together thoroughly, as was done with the other pastes. Put them in a crucible as usual, and cook them in a pottery furnace, or purpose-built kiln. They will make a deeper oriental sapphire of notable grace that has a tinge of violet.

~ 87 ~
Oriental Garnet

Have 2 oz of rock crystal ground impalpably as usual. Add 6 oz of ordinary minium, making 8 oz in all. Mix these powders thoroughly, and to each ounce of material give a proportion of 2 grains per ounce of prepared Piedmont manganese. In all, there will be 16 grains [of manganese]. Mix in 3 grains of prepared zaffer. Put this powder in a crucible that will resist the fire, and lute it as usual. Cook it in a furnace or purpose-built kiln as usual. It will make a most beautiful and sightly garnet.

~ 88 ~
A Deeper Oriental Garnet

Take 2 oz of ground rock crystal, as before, and 5½ oz of ordinary minium. Mix them together, and to each ounce of material, add 3 grains of prepared Piedmont manganese. In all, there will be 15 grains [of manganese]. Blend these thoroughly and mix in 4 grains of prepared zaffer. Put it in a fire resistant crucible, which has a space of more than a finger at the top, because this material boils more than the others do. Lute it as usual, like the other crucibles, and then cook this paste in a pottery furnace, or purpose-built kiln. You will have an oriental garnet color with a most beautiful violet tint.

~ 89 ~
Another Beautiful Garnet

Grind 2 oz of rock crystal impalpably as usual. Add 5 oz of ordinary minium. To each ounce of this material, give a proportion of 4 grains of prepared Piedmont manganese, as described, which will be 1 pwt and 4 grains [of manganese] in all. Into this, mix 6 grains of prepared zaffer. Blend all these things together well, and place them in a crucible that will resist the fire. Leave a space of more than a finger's width at the top, since this material boils, and swells quite a bit. Lute it well, let it dry, and then put it in a pottery furnace to cook, or else in a purpose-built kiln. It will become the color of true oriental garnet, and be more beautiful and sightly than any of the other garnets.

~ 90 ~
Advice about Pastes, and Their Colors

Be aware that the colors for the pastes described above can be made more or less saturated according to your will, and inclination, and according to the works in which they will be used. In order to make small stones for rings, the color must be deeper, for larger stones use less color, for earrings, and pendants, concentrate the color. Remit this matter to the discretion of whoever is doing the work. There are no true rules here; the rules given by me above serve only to illuminate the intellect of the curious artisan.

To this end, you can always invent and find better colors than the ones that I describe here, other than verdigris, zaffer, and manganese. The curious and noble spirit will be able to get a marvelous red from gold, yet another beautiful red from iron, a most beautiful green from copper, a golden yellow color from lead, a celestial one from silver, and a sky blue even more beautiful from the [red] garnets of Bohemia. You will find the small garnets inexpensive, and you can extract their tincture, as I have done many times in Flanders.

These materials result in notable effects, and you may do the same with rubies, sapphires, and other gems. All of these things are manageable by a practical chemist. To require me to write such things here would be too much of a tedious matter, especially since I wish to speak succinctly in the present work. However, with the colors described above, you can make very appealing pieces.

Because it is such a significant matter, I will repeat once more, that when the pastes are not cooked and purified enough, you should return them to cook again in the same crucibles, taking care not to break them. If the pastes are not satisfactorily cooked and purified, putting them into other crucibles will cause many impurities to form. They will have adhesions from the crucibles, and be loaded with dirt; they will be useless, and not usable for any work. When they are not cooked and purified well, if the crucibles are not broken in any way, but re-luted, and returned to the furnace or kiln to re-cook, then they will become pure, and beautiful, to be used in any kind of work that you might want to make.

~ 91 ~
A Marvelous Way, No Longer Used, To Make the Above Paste, and to Imitate Every Type of Gem

I took this method from [the writings of] Isaac Hollandus while I was in Flanders. As far as I know, no one uses it any longer to imitate gems. Known to only a few persons by chance, it is quite laborious, yet as large a production as it is, it makes gems of as much or more beauty and grace as perhaps any made anywhere until this day, or at least any shown to me by anyone. I will show this method most clearly, and with so many details, and warnings that one practiced in chemistry will easily be capable, and do the work perfectly.

Take ceruse of lead otherwise called biacca, and grind it most finely. Put it into a large glass urinal, and pour a lot of distilled vinegar over it, covering it to about the depth of one palm. Proceed carefully, because in the beginning the vinegar bubbles, and swells strongly. Therefore, add it little by little, slowly, until the fury and rumbling passes.

Afterwards keep this urinal in sand over a hot furnace, until 1/8 part of the vinegar evaporates. Remove it from the fire, and when the urinal cools, gently decant off the strongly colored vinegar, saturated with salt [lead acetate]. Put this part aside in a glass vessel, and return the residue of the biacca with new distilled vinegar to a low fire, as before, to evaporate 1/8 part.

Decant the colored vinegar as before, and repeat this operation, with the distilled vinegar many times, until you have extracted all the salt from the ceruse. This will be when the vinegar is no longer colored, and no longer has a sweet taste [☠], which usually happens by approximately the sixth time, at which point these colored vinegars are filtered together with the usual diligence.

Evaporate the filtered liquid from a glass urinal, leaving the salt of lead [lead acetate] dried in the bottom; it will be white in color. Pack this into a luted glass flask and put it in sand. Fully cover it in sand from the neck down, except for the mouth of the flask: leave it opened to the furnace to admit enough heat. Continue to heat the flask for 24 hours.

Then remove the salt from the flask, and grind it. If it is red [minium] like cinnabar then do not return it to the fire, but if you see it is yellowish [litharge] then return it to the fire in the glass vessel, as before, for another 24 hours so that it will become red like cinnabar. Have a good fire but not hot enough to melt it, because then all the hard work and the material will be lost.

Put this calcined red lead into a glass urinal, and as before add distilled vinegar. Repeat the work as before, in all and for all, until you have again extracted the salt from the dregs and sediment in all or in part. Hold these colored vinegars in glazed earthen trays to settle for 6 days, so that all sediment and imperfection will go to the bottom. Then slowly filter it, the majority of material in the bottom being unusable.

Now separate this well-filtered vinegar from any sediment and leave it uncovered [to evaporate] in a urinal. In the bottom the intensely white salt of Saturn will be left, which is as sweet as sugar [☠]. Dry it well and dissolve it in common water. Leave this solution in pans for 6 days, so that the sediment will go to the bottom. Now filter this salt saturated water, separate it from its unusable larger part, and evaporate it in a glass urinal.

> 80 LIBRO QVINTO
> pra l'altro colorito, & questa operatione, con l'aceto destillato si reiteri tante uolte, che habbino cauato dalla cerussa tutto il suo sale, che sarà quando li aceti non saranno più coloriti, & non haranno più gusto di dolcezza, che suole succedere alla sesta volta in circa allora questi aceti colorati, & insieme vniti si feltrino con la diligentia ordinaria, & feltrati si isuaporino in orinale di vetro, & si asciughino, che in fondo sarà il sale di piombo in colore bianco questo in boccia di uetro stacciata come vn leuto, lutata si tenga in Arena, che la boccia sia tutta drento nella Rena dal collo in fuora, la bocca della boccia si lassi aperta il fornello uadi assai caldo, & si continui per ventiquattro hore , poi si caui la boccia, & si macini questo sale, & se è rosso come cinabro non si ritorni più in fuoco, ma non sendo cosi ma giallignolo si ritorni in fuoco in vaso di vetro, come sopra per altre ventiquattro hore, che verrà rosso come vn cinabro, habbi buon fuoco pero non fonda, che saria persa tutta la fatica, & l'opera. Questo cosi rosso piombo calcinato si metta in orinale di vetro, & per sopra si metta aceto destillato, reiterando l'opera di sopra in tutto, & per tutto, fino habbi di nuouo cauato tutto il sale, & le feccie, & terrestreità in tutto ò parte questi aceti colorati tienli in catinelle di terra inuetriate per sei giorni, che ogni terie-streità, & imperfettione anderà in fondo allora si feltrino lassando la parte grossa in fondo come mutile, allora questi aceti ben feltrati, & separati da ogni terrestreità si scroprino in orinale, nel fondo del quale rimarra il sale di Saturno bianchissimo, e dolcie come zucchero, quale bene asciutto, si solue in acqua comune, & soluto si lassi in catinelle e per sei giorni, che sarà in fondo la terrestreità, al ora si feltri questa acqua pregna di sale, & separata dalle parte grosse, & inutile, si isuapori in orinale di vetro, che rimarrà in fondo vn sale bianchissimo quanto la neue, & dolce quanto il zuchero, reiterando però il soluere, & feltrare, & suaporare con l'acqua comune per tre uolte, allora questo si domanda zucchero di Saturno quale in boccia di vetro, o palla si tenga a calcinare in Arena in fornello, che habbi fuoco temperato per più giorni, che verrà calcinato in vno colore rubicondissimo più del Cinabro, & sottolissimo impalpabile
> come

Left in the bottom will be a salt as white as snow, and as sweet as sugar [☠]. Repeat the dissolution, and filtering, and evaporation with common water three times. This is the required sugar of Saturn. Keep it to calcine in sand in a glass flask or ball in a furnace over a moderated fire for many days. It will further calcine to a color that is much redder than cinnabar, and more finely impalpable than sifted grain flour. This is the required true sulfur of Saturn; purified from the sediment, foulness, and blackness that were upon the lead at first.

Now when you want to make pastes for emerald, sapphire, garnet, for topaz, chrysolite, or for a celestial, or other color, use the same materials, colorants, and doses, that are described above in the other prescriptions. Except, take the present sulfur of Saturn instead of using ordinary minium. Use the same doses, and colorants as have been described above for the other colors.

Always work most punctually, as before. You will have jewels of marvelous beauty in every color, which by far surpass those described above, made with ordinary minium. Because with this true sulfur of Saturn, they will surpass all others by far more than I can write here, as I have seen and made many times in Antwerp. The pastes made with this sulfur do not have the unctuousness, or yellowishness, that they ordinarily have [with minium], which in time cause them to become ugly. While some blemish quite a bit by perspiration and human sweat, this does not happen with these. Therefore, your tedium is not as great, because of the ample compensation of the work, and effect.

~ 92 ~
A Way to Make Hard Pastes in All the Colors

Have rock crystal, calcined and ground in a mortar, as was described distinctly in its place [chapter 76]. As an example use 10 lbs of this, and 6 lbs of Levantine polverino salt, extracted in a glass urinal, and very well purified, as was described in its place in the first book in the third chapter.

Dry this salt well and grind it over a porphyry stone. Pass it all through a sieve, and mix it in thoroughly with the above crystal. Then have 2 lbs of sulfur of Saturn made chemically, as was described in its place [chapter 91]. Unify these three powders very well in a clean glazed earthen pan, and make it into a paste with clean common water. Use only a little water so the paste is hard.

Make small cakes of 3 oz, and make a hole in the middle of each. Dry them in the sun, then move them to a pottery furnace to calcine, in the upper part of the furnace, or other similar fire. Then break up these small cakes, and grind them over porphyry stone thoroughly, passing the material through a fine sieve. Then move this into crucibles in the furnace and leave the glass to clarify for 3 days. Cast it into water, and then reheat it, and leave it to clarify thoroughly for 15 days. At this point, all the imperfections and the bubbles will be gone.

What remains is a most pure paste, like the natural gems, and furthermore this most pure type of glass will tint in any of the colors that you want. As an example use thrice cooked copper for an emerald color, as was done in the ordinary glass; also use the copper for an aquamarine; use zaffer for topaz; and use zaffer and manganese for sapphire. As with the ordinary glass, add the tincture once the glass has clarified and with the same diligence. You can make yellow with wine dregs and manganese adding them in parts, or even a garnet color with manganese and zaffer added in parts.

Make the glass as usual, and the effect will imitate all the gems in all the colors. This glass paste is most fine and will be quite hard, and have a shine, and polish of great wonderment. It closely imitates the natural stones and is nearly the same in hardness, especially the emerald; with this method, it will be most beautiful and almost as hard as the natural stones.

The Sixth Book of The Art of Glass

BY PRIEST ANTONIO NERI

In which I show the way to make all of the goldsmiths' enamels to fire over gold in various colors. Included are the rules, the colorant materials, and the methods to make the fires for such enamels with exquisite diligence. I present the demonstrations as clearly as possible for this subject.

Shown in this sixth book are the ways to make enamels of many colors for use by goldsmiths to enamel over gold and other metals with taste and grace. Now, this is not only a difficult art but also necessary. We see ornate enameled metals in many colors, and they make a pleasant and noble sight; they entice others to look and take notice.

In addition, enameling is one of the main segments of the glass field, and quite necessary. It seems to me to cause universal gratitude and pleasure, so I will endeavor to describe many ways to make all sorts of enamels, which are special materials in the art of glassmaking. They form one of its noble domains, not common, but a particular niche, and because this work is not lacking in substance, and is pleasant, useful, and necessary, I have made this present sixth book, for the satisfaction and benefit of everyone.

~ 93 ~
The Material to Make All Enamels

Take for example 30 lbs of fine lead, and 33 lbs of fine tin. Blend these metals together and calcine them in the kiln, as was described for lead in its place [chapter 62], and then pass this calx through a sieve. Boil the calx in clear water, in a clean earthenware vessel, which is to say in a kettle. When it has boiled a little, lift it from the fire and empty the water per your inclination, so that it will carry away the finer metallic calx. Replace it with new water over the sediment of the calx, and return it to boil. Decant this as above, and repeat many times. Do this so the water does not carry with it the bulk of the calx, so that the larger sediment will remain in the bottom.

In order to get finer pastes, return the calx to calcine and then boil it in common water, as above. Then evaporate all the water that has carried with it the fine part of the calx. Do this on a low fire, especially near the end so as not to spoil the calx. The finest calx will remain in the bottom, much more so than usual.

Take 50 lbs of this fine calx, and 50 lbs of crystal frit made with thoroughly ground white tarso. Through a sieve, sift 8 oz of white tartar salt, as previously instructed. Thoroughly pulverize everything, and mix it together, pass it through a sieve, and set this material down in the terracotta pot again, giving it fire for 10 hours. Remove the material, pulverize it well, and store it in a dry place. Cover it so dust does not get inside, since this is the [base] material to make all enamels, in all the colors.

~ 94 ~
Milk White Enamel [Lattimo]

As an example, take 6 lbs of enamel base material [chapter 93] and mix it with 2 pwt of prepared Piedmont manganese. Put this material in a crucible glazed with white glass, and into a furnace, with a clear smokeless fire using oak wood. Leave this material to melt and clarify. As soon as the work is well fused and consolidated, throw it into clear water, and then return it to the crucible. When it fuses and clarifies, again throw it into water, and return it to the crucible. When it is fused and clarified throw it in water, and return it to the crucible to fuse and clarify a third time.

Throw it into water, return it to the crucible, and leave it to clarify thoroughly. Take out a proof, seeing if it is as white as it should be. If it is greenish, give it a little manganese, as was shown previously. It will become milk white, which is tried and true for enameling over gold and other metals, as the goldsmiths do.

~ 95 ~
[Two] Turquoise Enamels

[Greenish]
Take 6 lbs of enamel base material. Put it into a crucible glazed with white glass, and leave it to melt and to clarify. When it is well fused and clarified, throw it into water again, return it to the crucible, and leave it to melt and clarify. Now give it 3 oz of thrice cooked copper, made in the furnace, as described in its place [chapter 28], and 4 pwt of prepared zaffer. Mix these two powders well.

You should add the material in four doses, stirring [the melt] thoroughly each time, and leaving the glass to incorporate the powder. The way to check that the color pleases you is by proofing it, and watching to see if it is sufficiently loaded. Stop adding powder and proceed to make a goldsmith's proof [on metal]; always examine the colors to get to know them by eye, as I have always done, because in this matter, I cannot give specific doses. Sometimes the powder will tint more, other times less, therefore you must practice with your eyes to understand the colors.

The way to learn about a color, when you need to make extra, is to add more enamel base material, which will quickly clarify and reduce the color. Then after reducing the color, add more powder, which tints. Thus, you will reduce and increase it such that it returns to the same color. At this point, take the turquoise [enamel] from the furnace, and out of the crucible. As is usual for enamels, make dollops of about 5 - 6 oz each. This will produce very beautiful and sightly enamel for goldsmiths.

[Bluish]

Into a crucible glazed with white glass, in a glassmaking furnace, put 6 lbs of enamel base material. To this, add a thorough mixture of 3 oz of prepared zaffer, blended with 2 pwt of prepared Piedmont manganese. Mix and unify these two powders thoroughly with the above [base] material. Once mixed, throw them into the crucible, where it will soon clarify. Clean this material by pouring it into water, and then reheat it; watch to see if the color appeals to you, and if the tint is sufficient.

Now you can either intensify, or lighten the material by concentrating, or diluting the colors. When the color is too strong, add to the fused [base] material. Too much of this material will make enamels without enough color. So reduce the excess coloration in small amounts, little by little, until it reaches the desired tint.

When the color is too weak, give it more of the powder, coaxing [the color] little by little to fruition. Always make sure to test it from time to time. This is the way to add all the colors, because this way you will never fail. In Pisa, I made them without [measuring] weights, but by rough estimate.

I have colored every type of glass for every job, enough to elucidate this method; I leave the rest to the curious observer, the ingenious worker, and the artist. Leave this material alone in the crucible until it is well cooked, and the color is thoroughly incorporated. Then take it out of the furnace as usual. This azure blue enamel will be one of the goldsmiths' most beautiful.

~ 96 ~
Another Blue Enamel

As an example, take 4 lbs of enamel base material. With this, thoroughly mix 2 oz of prepared zaffer, but first mix in 2 pwt of thrice cooked copper, made with the flakes from the kettle-smiths, as was described in its place [chapter 28]. Thoroughly blend these two [unified] powders with above enamel base material.

Once they are mixed, throw them into a crucible glazed with white glass, and put them in the glass furnace. When it is fused and well clarified, dump it into water, and repeat. Put it into the furnace and leave it to cook and clarify thoroughly. Finally, remove it from the furnace. The result will be a most sightly blue enamel for goldsmiths.

~ 97 ~
Green Enamel

Into a crucible glazed with white glass as usual, put 4 lbs of the above [chapter 93] enamel base material. After 10 - 12 hours in the furnace, it will fuse and clarify thoroughly. Throw it in water, and reheat it in its crucible. Leave it to clarify thoroughly.

When clarified, give it 2 oz of thrice cooked copper made with sheets of copper in furnace trays, as described in its place [chapter 25]. Mix in 2 pwt of well ground iron flake, which falls from the anvil. Add these two powders to the material in three doses, always mixing the material well so that the color will incorporate. Watch for when the color appeals to you. When you see the sign that it is right, leave it to clarify well, and fully incorporate the color. Remove it from the furnace as usual. You will have a beautiful green enamel for goldsmiths.

~ 98 ~
Another Green Enamel

As an example, start with 6 lbs of enamel base material. With it, thoroughly mix in 3 oz of well ground Spanish ferretto. Into this, mix 2 oz of crocus martis. Once these materials are mixed and blended well, put them in a crucible glazed with white glass and leave it [in the furnace] to clarify thoroughly.

Once clarified throw it into water, and return it to the crucible to clarify again. Watch to see if the color appeals to you, so you can concentrate, or dilute it. When you see signs that the color is right, leave it to clarify, and to incorporate the color. Take it out of the furnace as usual. You will have a beautiful green Goldsmith's enamel. Make these furnace enamels into dollops approximately 4 - 6 oz each.

~ 99 ~
Another Green Enamel

Into a glazed furnace crucible put 4 lbs of the above enamel base material [chapter 93], and melt it as usual. In a short time, it will clarify. Now throw it in water, and return it to the crucible. Leave it to clarify again. Now mix the following two powders together and add them in three doses. That is, 2 oz of thrice cooked copper made from flakes of kettle-smith's hammerings, as described in its place [chapter 28], and 2 pwt of crocus martis made with vinegar [chapter 17].

Mix these powders thoroughly and give them to the above material when it has fused and fully clarified. Mix and incorporate the powder taking care to watch closely, and to watch for signs that the color is right. As such, remove it as usual and form it into cakes, but first allow it to clarify, and fully incorporate the color. This will be a beautiful and graceful enamel for goldsmiths.

~ 100 ~
Black Enamel

Have as an example, 4 lbs of the above said [chapter 93] enamel base material, and 4 oz of the previously described powder, that is to say 2 oz of prepared zaffer, and 2 oz of Piedmont manganese. Mix these two powders thoroughly and unify them with the above base material.

Once mixed, put them into a crucible glazed with white glass, and leave them to clarify. This material swells quite a bit; therefore, the crucible must be larger so the material will not erupt out. Once it has clarified, throw it into water, and then return it to the crucible, where it will clarify quickly. Watch to see if the color is satisfactory, or should be made weaker, or stronger according to your needs. When the enamel is ready, you should make dollops of it as before. It will be a very beautiful velvet black goldsmith's enamel.

~ 101 ~
Another Black Enamel

Have 6 lbs of the enamel base material described above [chapter 93]. Now take 2 oz of prepared zaffer, 2 oz of crocus martis made with vinegar [chapter 17], and 2 oz of Spanish ferretto. Grind these three powders well, blend, and mix them thoroughly with the above said base material.

Inflame this mixture in a furnace crucible glazed as usual. When it has fused and clarified, throw it into water, and then return it to the crucible, where it will purify quickly. Watch to see if the color develops to your liking, remove it, and form it into cakes as usual. If you clarify the glass well, and incorporate the color well, this will be a beautiful black enamel for goldsmiths.

~ 102 ~
Another Black Enamel

Have 4 lbs of the above described [chapter 93] enamel base

material, 4 oz of tartar, or else gruma from bottles of red wine, and 2 oz of prepared Piedmont manganese. Grind these powders well, and blend them together. Mix them thoroughly with the above base material. Put it into a furnace crucible glazed with white glass.

Make sure that the crucible is large with a lot of the space, because this material will swell quite a bit. Let it melt, and clarify well, then throw it into water, and return it to the crucible. Let it clarify again, watching to see if the color appeals to you. When the signs are right, remove it as usual and form it into cakes. This will be the most beautiful velvet black for goldsmiths to enamel over metals as usual.

~ 103 ~
Red Wine Colored Enamel

To 4 lbs of enamel base material [chapter 93], give 2 oz of well-prepared Piedmont manganese. Mix them thoroughly, and then put this material into a furnace in a crucible glazed as usual, but in a very large [crucible] so there remains a lot of space. This material boils and swells up quite a bit.

When it fuses well and clarifies, throw it into water, return it to the crucible, and leave it to clarify again. Watch to see if the color is to your discretion. Once the signs are right, remove it as usual, and form it into cakes. This very beautiful enamel will be a purplish color, most beautiful for enameling. Add more or less of the color according to your need.

~ 104 ~
Purple Enamel

For example, start with 6 lbs of enamel base material, as described above [chapter 93]. With this, thoroughly mix the previously described powder, that is to say 3 oz of prepared Piedmont manganese, and 6 oz of thrice cooked copper made with flakes of the kettle-smiths, as was shown [chapter 28].

Unite these powders, and mix them well with the above said base material, which goes into the furnace, in a crucible glazed as usual. Leave it to clarify well, then throw it into water and return it to the same crucible. Leave it to cook and to clarify. See if the color is to your discretion. You can concentrate, or dilute the color, as required. Remove it as usual, form it into cakes for use by the goldsmiths. It will be a beautiful enamel.

~ 105 ~
Yellow Enamel

Take 6 lbs of enamel base material, and 3 oz of gruma from bottles of red wine, and 3 pwt prepared manganese. Grind these powders well, and blend them together. Mix them thoroughly with the above said base material. Once mixed, put them into a furnace in a crucible glazed with white glass. It should be rather large; this material will swell a lot, and should not overflow the crucible.

Once clarified, throw it into water, and then return it to

reheat. Leave it to clarify again, watching to see if the color appeals to you. If not, you can concentrate, or to dilute it according to your needs. Then remove it and form it into cakes. This will be a beautiful yellow enamel with which to color the metal gold in particular. However, it will not look as good as what the goldsmiths make when put over enamels of other colors.

~ 106 ~
A Celestial Blue Enamel

Use 2 oz of calcined tinsel and 2 pwt of prepared zaffer in order to make the celestial color of the blue magpie. As has been taught, mix these powders well and unify them with 4 lbs of enamel base material [and melt].

When it clarifies, throw it in water as usual, and then return it to the crucible. Again, leave it to melt, and clarify thoroughly. Now, check to see if the color is right, so it can be concentrated or diluted, according to your need, as has been shown for the other enamels. When signs show the color is right, take it out as cakes as described previously for use by the goldsmiths. This will be a very graceful and sightly celestial color.

~ 107 ~
A Violet Enamel

If you want to make a violet enamel that is both beautiful and handsome, start with 6 lbs of enamel base material [chapter 93]. Take 3 oz of prepared Piedmont manganese, and 2 pwt of thrice cooked copper made from the hammerings of kettle-smiths [chapter 28].

Mix these two powders together thoroughly, and unify them with the above said material. Put it into a crucible in a furnace as usual. Throw it into water, and reheat. If signs show the color is right, and it is not necessary to concentrate it or to dilute it, remove it in cakes. This will be beautiful enamel for goldsmiths in its color.

[blank page]

This gold powder, given in proportion little by little,
will make the transparent red rubino glass;
but you must experiment in order to find it.

- Antonio Neri chap. 129

The Seventh Book of The Art of Glass

By PRIEST ANTONIO NERI

In which I show the way to extract a yellow lake for painters from broom flowers, and all the other colors. With another way to extract lakes of red, green, blue, purple, and of all the colors from all types of plants and flowers.

I show the way to make ultramarine blue and alemagna blue, and a way of extracting lake from kermes, the verzino tree, and madder root, for paint. A method describes how to re-color faded turquoise. I also show another way to make transparent [red], and ways to make rosichiero for enameling over gold and other metals; these are things not ordinary or common.

This present last book shows the way to extract all the colors from flowers and plants for use by the painters, which are perfectly good for the glasses. The way to extract lakes of many colors, and the way to extract ultramarine blue from lapis lazuli well; these are all things of particular use by

---

LIBRO SETTIMO
DELL'ARTE VETRARIA
DI PRETE ANTONIO NERI

Nel quale si mostra il modo di cauare la Laccha gialla per Dipintori, da fiori di Ginestra, & da tutti li altri colori; Con vn altro modo da cauar la Laccha rossa, Verde, Azzurra, Pagonazza, & di tutti, i colori da ogni sorte di Erbe, & fiori. & il

MODO di fare l'Azzuro di Alemagna, & l'Azzuro oltramarino, con il modo di cauar la Laccha dal Chermisi, dal zino, dalla Robbia per di pintori, il modo di colorire le Turchine, scolorite; Vn'altro di fare l'osso trasparente, & il rosichiero per smaltare sopra l'Oro, & altri metalli, cose non vulgari, ne comuni. Nel presente vltimo libro mostro il modo di cauare tutti i colori da fiori, & erbe per vso de i Dipintori, che possono seruire ancora per i vetri; Il modo di cauare le Lacche da molti colori; il modo di caare l'Azurro Oltramarino da Lapislazzuli, cose tutte se bene in particolare per vso de i Pittori, possono nondimeno seruire nell'arte del vetro per dargli colore, & in superficie, & anco in sieme nel fuoco, come è l'Azzurro Oltramarino, mostro il modo di fare il rosso trasparente in vetro, che hoggi pare sia del tutto spento cosa non vtile; Il modo di fare il Rosicchiero da smaltare sopra Oro, tutte

painters, but nevertheless serve in the art of glass to give it color, both on the surface, and also to inflame in the fire, like the ultramarine blue.

Also shown is the way to make transparent red glass, and other things that today seem obsolete and disused, like the way to make rosichiero for enameling over gold. All stuff of the glassmaking art, and today quite obscure, and seldom noted [are shown] along with many other things that I have judged must go into the present work, which I believe will gratify the curious and noble spirit.

~ 108 ~
A Yellow Lake from Broom Flowers, for Paint

Make a batch of lye from glassmaking soda and lime, and make it reasonably strong. In this lye, boil fresh broom flowers over a slow fire until the lye draws out all the tincture of the flowers. You will know this [state] when you take the flowers out and see that they have turned white, and become thoroughly uncolored, and the lye is as yellow as a fine Trebbiano wine.

Now take out all the flowers, and put the lye in glazed earthen pots over the fire, so the lye boils. Into this add as much roche alum as can be dissolved with the fire, then remove it, and empty the lye into a vessel of clear water. The yellow color will settle to the bottom. Leave it to rest, and then decant off all the water. Again, pour more fresh water over it, decant again, and let it rest. As before, the dye will go to the bottom. This way, you will extract all the lye salts and the dose of alum out of the dye.

Take note that the more you wash this dye of the lye salts and alum, the more beautiful it will become, and more lovely in color. Always wash it with common water, which will carry away the lye salts and the alum, hand in hand.

First, decant the water, and then leave it until the yellow dye settles to the bottom. Repeat this until the water tastes sweet without saltiness. When you notice this, it will be the sign that the water has washed away all the salt and alum. The pure and beautiful lake will remain in the bottom. Drain it well from the water, and while still damp spread it over pieces of white linen, and leave it to dry over newly baked tiles. Let it dry in the shade, and you will have a beautiful yellow colored lake for the painters and for glass as well.

~ 109 ~
A Way to Extract Lake from Poppies, Blue Irises, Red Roses, Violet Roses,

> DELL'ARTE VETRARIA   95
>
> *Al cauare la Laccha di Rofolacci, Fioralifi, Rofe Roße, Viole roffe, & da ogni forte di Erba verde.   Cap CIX.*
>
> HABBI quella quantità di foglie di fiori, che uorrai, in qual fi fia colore, però ogni colore da per fe, ouero herbe verde belle pur da per fe, & quefte materie in lifcia forte fatta di foda, & calcina come fi dice di fopra nella Lacca dei fiori di gineftra cauandone la tintura, & dalli allume, & acqua frefca facendola dare in fondo in tutto, & per tutto, come fi dice nella Lacca de i fiori di Gineftra, & con acque frefche lauandole a più acque da ogni falfedine, & allume, & da vltimo afciugandola in pezze di panno lino, come in quefta maniera auerai la Lacca, & vero colore, e tintura da ogni fiore, & herba, che per Pittori farà cofa vaga, & bella, & fenza dubbio degna di effere ftimata affai.
>
> *A cauare la Lacca, & colore per dipingere da fior' Ranci, Rofolacci, Fioralifi, Viuole ordinarie, Viuole roffe, Rofe incarnate, Rofe roffe, Fiori di Borrana, Fiori di cappucci, Fiori di ghiaggi uolo, & da ogni fiore di qual fi voglia colore, & il verde della Malua, della Pimpinella, & di tutte l'Erbe. Cap. CX.*
>
> PIGLISI qual fi uoglia fiore di qual fi uoglia colore, o uero herba, che ftropicciata verde fopra il foglio, o carta bianca la tinga del fuo colore, che fara buona, perche l'herbe, & fiori, che non fanno quefto effetto, non fono buone. Adunque in vn'Orinale di vetro fi metta acqua vite ordinaria, & nel fuo cappello, auuertendo che il criftallo di detto cappello fia largo il più che fia poffibile, & in detto rifalto fi mettino le foglie diquel fiore, o herba dalla quale fi vuole efuberare, & eftrarne la fua tintura, poi fi luti le giunture del cappello, & fi adatti al fuo roftro il recipiente lutate le giunture, fi dia fuoco temperato che la parte fottile dell'acqua uite afcendendo nel cappello, & cafcando nel rifalto di effo cada addoffo alle foglie di fiori, efubera la tintura, & cafca dal roftro del capello nel recipiente colorita, & carica della tintura del hore, facendo paffare tutta la parte fottile de l'acqua vite, poi viene colorita, quefta
> parte

and From All Kinds of Green Plants

Have whatever amount of flower petals you want, in which there is color, but keep each color separate. You can use pretty green plants as well, also by themselves. Put these materials into strong lye made of soda and lime as described above for the broom flowers lake recipe [chapter 108]. Extract the dye, and give it alum, and fresh water making [the pigment] settle in the bottom. Do this, in all and for all, as was described in the broom flower recipe.

With more fresh water wash away all the saltiness and alum, and finally dry it on pieces of linen cloth. In this way realize the lake, and true color, and tincture from every flower, and plant. For painters it will be a thing of grace and beauty, and without a doubt worthy of high regard.

~ 110 ~
A Way to Extract the Lake and Color for Painting, from Orange Blossoms, Red Poppies, Blue Irises, Ordinary Violets, Red Violets, Carnations, Red Roses, Borage Flowers, Day Lilies, Irises, and From Flowers of Any Desired Color, and the Greens of the Mallow, the Pimpernel, and All the Plants

Take whichever flower you want, of any color you want, or even a [green] plant. If it will rub green from a leaf onto white paper staining it with color, then it will be good. The plants and flowers that do not show this effect are no-good. Put ordinary grappa into a glass urinal, with an alembic for its cover, making sure the said crystal alembic is as wide as possible.

Into this alembic pack the leaves [or petals] of any flower or plant from which you want to release and extract the tincture. Now lute the mouth joint of the alembic. Fit a receiver to its snout, and lute that joint. Give it a moderated fire so that the volatile part [alcohol] of the grappa rises into the alembic, and falls down into its volume upon the petals of the flowers, extracting the tincture.

In time, drops will run down the snout of the alembic into the receiver, colored and charged with the tincture. Once all of the volatile part of the grappa passes, and becomes colored, distill this colored volatile part of the grappa in a glass vessel. [The alcohol] will pass white, and will be useable three more times. The dye will remain in the bottom, which you should not allow to dry too much, but just moderately. Then you will have the very best tincture or lake for painting from an abundance of flowers and plants.

~ 111 ~
An Azure [Paint] like Alemagna Blue

Take 2 parts mercury [by weight], and then grind 3 parts flowers of sulfur with 8 parts sal ammoniac, both over a porphyry stone. Add [the powder] to the mercury in a glass ball with a long neck, luted on the bottom. Set it in sand and give it a slow fire [☠] to drive out the moisture. Then stopper the mouth of the glass, and build the fire, as is done for sublimation, continuing the fire until it burns out. You will have a quite beautiful and graceful blue.

~ 112 ~
The Way to Color Faded Natural Turquoise

Take turquoise that has faded and turned pale, put it into a glass ampoule, and cover it over with oil of sweet almonds. Keep this ampoule lukewarm in ash over a moderated [heat]. In 2 days at the most, it will have acquired a most beautiful color.

~ 113 ~
A Mixture to Make [Mirror] Spheres

Have 3 lbs of well-purified tin, and 1 lb of copper also purified. Melt these two metals, first the copper, then the tin. When they fuse thoroughly, throw onto them 6 oz of just singed red wine tartar, and 1½ oz of saltpeter, then ¼ oz of alum, and 2 oz of arsenic. Leave these all to vaporize, and then cast [the metal] into the form of a sphere. You will have good material, which when you burnish and polish, will look most fine. This mixture is called acciaio [white bronze] and is used to make spherical mirrors.

~ 114 ~
The Way to Tint Glass Balls, and Others Vessels of Clear Glass, From the Inside, In All Kinds of Colors, So They Will Imitate Natural Stones

> DELL'ARTE VETRARIA 97
>
> *Modo di tingere Palle di vetro, o altri vasi di vetro bianco, per di drento, d'ogni sorte di colori, che imiteranno le pietre naturali.* Cap. CXIIII.
>
> HABBI la palla di vetro, ouero altra sorte di vetro, che sia bianco, e bello, & piglia colla di pescie, che stia infusione in acqua comune per dua giorni di poi piglia di detta colla di pescie stata infusione, & mettila in vno pignattino con acqua chiara, & fa bollire fino si stemperi benissimo tutta, auuertendo che la colla vuole essere tenerissima con assai acqua, di poi leuala dal fuoco, & quando è tiepida mettine nella palla di vetro, & gira bene a torno il vaso, acciò la colla pigli, e bagni per tutto il vetro di dentro, fatto questo si scoli l'humidità, che escie, di poi habbia a ordine i colori poluerizzati, cioè minio, & gettane drento la palla di vetro spruzzando detto colore, che vadi a onde, con vna palettina fatta di canna buttato in più luoghi del minio, butta del smalto azzurro, spruzzando con detta paletta di canna a onde drento la palla a torno, poi farai l'istesso con verderame ben macinato, poi con Orpimento pur ben macinato, poi con Lacca ben macinata sempre a ogni colore gettando a onde in più luoghi, che mediante la colla, che harà bagnato la pasta drento, per tutto queste poluere si attacheranno al vetro, & così farai con tutti i colori, poi habbi gesso ben poluerizzato, & mettine assai nella pasta, & presto girala a torno, che si attacherà per tutto il vetro di dentro, facendo questa operarione presto, mentre l'vmidita della colla è fresca, acciò le poluere si attachino bene, poi uota per il buco della palla il gezzo che farà auanzato drento la palla, la quale per difuori apparirà colorita in diuersi colori con vna vista bellissima, che sembrerà scherzi naturali di pietra dura,& fine, questi colori, come la colla è asciutta bene si attaccano di maniera, che poi non si stachano più , e sepre per difuora il loro colore è belissimo a queste palle si adatta vn piede di legno, o di altra materia dipinto, & si tengono per bellezza sopra studioli, & in scrittoi con vista assai bella.
>
> N   *Azzuro*

Have a ball of glass, or else glass of another shape, that is clear and beautiful. Take isinglass [fish glue], that has been infused in common water for 2 days. Put this hydrated isinglass into a bowl of clear water, and boil it until it all thoroughly softens. Make sure there is enough water to make the glue quite soft, and then remove it from the fire.

When it is lukewarm put some in the glass ball, and swirl it around well. Turn the vessel, and in this manner bathe the entire inside of the glass with the glue. Then pour out the excess. Drain it and have the following colors ground and ready. Start with minium [red lead], pour it inside the ball of glass, sprinkling the color so that it runs in waves. Use a small spoon made of reed to cast the minium in more areas.

Next, throw in the blue enamel. Sprinkle it with the reed spoon forming waves [of color] within the ball. In turn do the same with well ground [green] verdigris, then with [yellow] orpiment also well ground, then with lake well ground.

Always for each color, throw it in waves, in new areas. By means of the glue, which will bathe the paste within, all these colored powders will adhere to the glass.

Now take well ground plaster of Paris, put some into the glass, and quickly turn it all about, so that it will adhere to the entire glass from within [backing the colors]. Do this operation quickly while the moisture of the glue is fresh, therefore the powders will adhere well. Empty the excess plaster inside through the hole in the ball. It will appear tinted in various colors in a most beautiful sight, which resembles natural hard stone toy amusements.

In the end, when the glue is fully dry, these colors affix themselves [to the glass] so they will never come loose. From the outside, the colors will always be beautiful. Affix these balls to wooden bases, or other painted materials, and keep them for their beauty on study shelves, and on desks, where they make a very beautiful sight.

~ 115 ~
Ultramarine Blue

Take fragments of lapis lazuli, which you can find plentifully in Venice, and at low prices. Get fragments that are nicely tinted a pretty celestial color and remove any poorly tinted fragments. Cull the nicely colored fragments into a pot, and put it amongst hot coals to calcine. When they are inflamed throw them in fresh water, and repeat this two times.

Then grind them on a porphyry stone most impalpably to become like sifted grain flour.

Then take equal amounts, 3 oz each, of pine pitch, black tar, mastic, new wax, and turpentine, add 1 oz each, of linseed oil, and frankincense. I put these things in a clay bowl to warm on the fire until I see them dissolve, and with a stirring rod, I mix and incorporate them thoroughly. This done, I throw them into fresh water, so they will combine into one mass for my needs.

For every pound of finely powdered lapis lazuli, ground as described above, take 10 oz of the above gum cake. In a bowl over a slow fire, melt the gum, and when it is well-liquified throw into it, little by little, the finely powdered lapis lazuli. Incorporate it thoroughly into the paste with a stirring rod.

Cast the hot incorporated material into a vessel of fresh water, and with hands bathed in linseed oil, form a round cake, proportionately round and tall. You should make one or more other of these cakes from the quantity of the material. Then soak these cakes for 15 days in a large vessel full of fresh water, changing the water every 2 days. In a kettle, you should boil clear common water, and put the cakes in a well-cleaned glazed earthen basin. Pour warm water over them, and then leave them until the water has cooled.

Empty out the water, and pour new warm water over them. When it has cooled, pour again, replenishing the warmth. Repeat this many times over, so that the cakes unbind from the heat of the water. Now add new warm water, and you will see that the water will take on a celestial color. Decant the water into a clean glazed pan, pour new [warm] water over the cake, and let it color [the water].

When it is colored, decant it and pass it through a sieve into a glazed basin. Pour warm water over the cake, repeatedly until it is no longer colored. Make sure that the water is not too

too hot, but only lukewarm because too much heat will cause the blue to darken, hence this warning, which is very important.

Pass all this colored water through a sieve into the basin. It still has the unctuosity of the gum, so leave it to stand and rest for 24 hours; all the color will go to the bottom. Then gently decant off the water with its unctuosity, pour clear water over it, and pass it through a fine sieve into a clean basin.

Pass the fresh water through the sieve with the color stirred-up so that this color still passes through, and therefore a great part of the filth and unctuosity will remain in the sieve. Wash the sieve well, and with new water again pass the color through. Repeat these steps three times, which ordinarily leaves all the filth on the blue resting in the sieve. Always wash the sieve each time, cleaning it of all contamination. Put the blue in a clean pan. Gently decant off the water, and then leave it to dry. You will have a most beautiful ultramarine, as I have made many times in Antwerp.

The amount per pound of lapis lazuli will vary. It depends on whether the lapis has more or less charge of color, and on the beauty of its color. Grind it exceedingly fine on the porphyry stone, as described above, and you will succeed beautifully.

For a quite beautiful and sightly biadetto blue that mimics ultramarine blue, take ordinary blue enamel. Grind it exceedingly fine over the porphyry stone, as above. Incorporate it into the gum cake with the dose described above, and hold it in digestion in fresh water for 15 days as with the lapis lazuli. Follow the directions for the lapis lazuli, in all and for all, until the end. These blues are not only useful to painters, but they also serve in order to tint glasses par excellence.

~ 116 ~
[Prepared Wool Shearings to Make a] Lake of Kermes for Painters

Take 1 lb of white woolen cloth shearings, which are of a fine wool, hold these shearings in fresh water for a day, then wring them out well, and this takes away the unctuosity that [the wool] has when it is sheared, which comes out of the hide. Now mordant them in the following way.

That is to say, take 4 oz of roche alum, and 2 oz of pulverized raw tartar. Put them in a small kettle with approximately 3 flasks of water. As soon as it begins to boil, put in the shearings and let them boil for ½ hour on a slow fire.

Then remove them from the fire, and leave them to cool for 6 hours. Now gather the shearings, wash them with clear water, and let them sit for 2 hours, then thoroughly wring the water out of the shearings and leave them to dry [in preparation for chapter 117].

~ 117 ~
An Elixir for extracting the color from Kermes

Put the following ingredients into a kettle: 4 flasks of fresh water, 4 lbs grain bran, ¼ oz of Levantine Saint John's wort, and ¼ oz of fenugreek. Leave it over the fire, so that the water becomes lukewarm, but so that you can still hold it with your hands. Remove it from the fire, and cover the kettle with cloth to preserve the color. Leave it for 24 hours, and then decant this lye or elixir for use.

Now take a clean kettle, and into it put 3 flasks of fresh water, and a flask of the above elixir. When it boils, put in [dried] kermes [bugs] crushed in the following manner: In a bronze mortar, grind 1 oz of kermes. Pass it through a sieve to ensure that you ground it well. Sift it many times, so that all the crushings pass through the sieve.

Finally take a little raw tartar, and crush it in the mortar. The tartar will take up all the dye adhering to the bottom of the mortar and to the pestle. Mix this tartar with the sifted kermes, and when the water in the kettle boils, put in all the kermes, and let it dye the water, reciting the Miserere Psalm once.

Then take the mordanted shearings, described above, which have first been soaked in a basin of fresh water for ½ hour. When the kermes tints the water [in the kettle] well, take the shearings, thoroughly wring out the [fresh] water, and throw them into the kettle. With a rod, stir the shearings in the kettle well, so that they take the dye thoroughly.

Leave it alone for ½ hour over the fire, boiling gently. Then remove the kettle from the fire. Collect the shearings by stirring them with a clean stick, and put them in a basin full of fresh water. At the end of ½ hour drain all the water, and add new fresh water. Wring them out well, and set them to dry in a place where dust will not fall on them. Spread them out so they will not get musty, and re-warm them. Make sure that the fire is always nice and slow, because with a strong fire the dye will turn black.

Afterward make some lye in this manner: Take the ash of vine twigs, or willows, or other softwood ash, put it onto a folded cloth, pour fresh water over it gently, and let it strain into a basin. Strain the water again through the ash two more times, and then let the lye rest for 24 hours, so that the ash falls to the bottom, and it is cleaned and well clarified. Then decant it into another basin, leaving behind the sediment and the part that is no good.

Take some of this lye and put it in a clean kettle, and while it is still cool put in the shearings tinted by kermes, and then bring it to a boil over a most temperate fire. This way, the lye will be tinted a red color, and will extricate the dye from the shearings. Now, take a little bit of the shearings, and wring them out well. If you find it is uncolored then remove the kettle from the fire. This is the sign that the lye has extracted all the dye of the kermes from the shearings.

Have a linen cloth stocking that is suspended over a very large basin, and through this stocking strain all the dye in the kettle. The shearings will remain trapped in the stocking. Once it drains, wring out the stocking with the shearings inside. Then get all the shearings by turning the stocking inside out and washing the fibers from it so it becomes neat and clean.

Now take 12 oz of pulverized roche alum. Put it into a large glass of fresh water, and leave it to stand so that all the alum dissolves in the water. When it fully dissolves, fit the linen stocking, well washed of the shearing fibers, over two sticks that are suspended in air, and very wide at the mouth, and constricted at the bottom. Sew this into the shape of an [inverted] cone. Keep a well-cleaned kettle under the stocking. Then take the glass of alum-saturated water, and pour it all into a basin containing the pigment of the kermes. Immediately you will see that the water will make the pigment of kermes separate, like a coagulant. Then pour all this dye through the stocking into the clean kettle, so that the lye strains clear through the stocking. The stocking will trap the dye of the kermes.

When all the liquid is well strained, if by chance it strains somewhat colored, then return it into the stocking. This time it will leave all the dye in the stocking. This second time the lye will be strained white, free of all dye, and the dye will all remain inside the stocking. Then take a clean small wooden scoop, and scrape the lye from the outside of the stocking, where it will cling grossly.

Have some ordinary newly fired tiles, and stretch linen rags over them. Over these rags, spread the lake that you remove from the stocking, and leave it to dry thoroughly. Do not spread it too thickly, because then it will not dry quickly. When there is too much moisture it will mildew, and make an ugly color.

When a tile has absorbed a lot of moisture, take another new tile. In this manner, it will dry more quickly. When it is dry, remove the coating from the linens. This will be a good lake for painters, as I have made many times in Pisa. Take note that if the color is too strong you should use more roche alum, and if it is too weak use less alum so that the color is according to your taste, and desire.

~ 118 ~
Very Beautiful Lakes of Brazil Wood, and Madder Root

If you want to extract a lake from either of these individual materials, do as instructed above [chapter 117] for the kermes, in all and for all. Tint the water with one of these materials, but do not give it as much alum per ounce as with the kermes.

Kermes has a more concentrated dye than Brazil wood, or madder. Therefore add it proportionately; with practice you will find that in this case, to each pound of shearings, you will add more Brazil wood or madder root, because they do not have as much dye, as the same amount of kermes. In this way, you will have a very beautiful lake for painters, with less expense than for the kermes. The [lake] of madder root in particular will be most beautiful, and a very sightly color.

~ 119 ~
Lake of Kermes another Easier Way

In this method, invented by me in Pisa, the shearings are not necessary, nor the elixir, the lye, or the dying of the wool, or many of the things done in the above description [chapters 116-117]. That way does work, but is quite laborious. This way is easier and it produces exactly the same results. Proceed in the following way:

Take the very best grappa, and into a flask of it put 1 lb of well-pulverized, fully fragmented roche alum. Then put in 1 oz of pulverized kermes, sifted in all and for all as above [chapter 117]. Add all this material to a glass flask with wide a neck. Agitate the vessel well, so that the grappa colors wonderfully.

Leave it alone for 4 days, and then empty this material in a clean glazed basin. Take 4 oz of roche alum, dissolve it in common water, and pour it over the basin of grappa tinted with kermes. Pour this into a stocking suspended over a basin, as was done for the other lake with the wool shearings. The grappa will strain through the stocking uncolored, leaving the dye in the stocking, and if it passes somewhat colored pour it through another time so that will pass clear.

Take the lake out of the stocking,

with a clean wooden scoop, and put it on pieces of linen to dry, stretched over tiles like the other lake, in all and for all, so that you will have a lake of kermes, most noble, with little hard work and in much greater amounts, all demonstrated by me in Pisa.

~ 120 ~
Transparent Red in Glass

Grind manganese impalpably, then mix it with an equal amount of refined saltpeter, and put it into a clay pan set to the fire, reverberating and calcining it for 24 hours. Take it then, and wash its saltiness away with warm common water. Once separated from the salt, let it dry. It will be a ruby-red color. With this, mix an equal weight of sal ammoniac, and grind them together over porphyry stone with distilled vinegar, which they will soak up. Leave this alone to dry, and then put it in a retort with a wide body and a long neck. Heat it in sand for 12 hours to sublimate.

Then break up the glass. Take all the deposits in the neck and body of the retort, and mix it with the residual remains in the bottom. Weigh it, and combine it all with as much sal ammoniac as was lost in the first sublimation. Grind everything together over the porphyry stone, with distilled vinegar for it to soak up. Then put it in a retort to sublimate as above. Repeat this sublimation, in this manner, many times until in the end, the manganese will all remain fusible in the bottom.

This is the medicine that tints crystal and pastes in a diaphanous red color, and in ruby red as well. Use 20 oz of this medicine per ounce of crystal or glass, but more or less may be used accordingly to govern the color. The manganese should be the very best from Piedmont, so that it will have the effect of tinting the glass a beautiful ruby color, and be a sight of wonderment.

~ 121 ~
Red like Blood

Put 6 lbs of lead glass and 10 lbs of common glass into a large crucible glazed with white glass. When the glass is cooked and clarified, throw some red copper into it. The way to make this I have taught [in chapter 24]. Use your discretion in adding it, and leave it to incorporate. Now stir the glass well and add the dregs of red wine, pulverized thoroughly, and the glass will become red like blood. If it does not color enough, add more red copper and wine dregs until it becomes this color. When [making] vessels, reheat [strike it] so it will form without a doubt.

~ 122 ~
Balas Color

Put crystal frit into a large crucible, [melt it] and throw it into water three times. Then tint it with prepared Piedmont manganese, so that it will become a nice clear purple, and add finely sifted potash. Put in enough of it so the glass becomes purple. Do this eight times, and know that the potash makes the glass turn yellowish, and somewhat reddish, but does not darken it, and always makes use of the manganese. The last time that you add the manganese, do not add any potash unless you colored it too much, and so you will have the most beautiful Balas color.

~ 123 ~
To Extract the Spirit of Saturn, Which Serves Many Uses in Enamels and Glasses

Take well ground litharge, put it in a glazed earthen bowl, and as above [chapter 91] add distilled vinegar, covering it to a depth of four fingers. Let it stand this way, until you see the vinegar tint up in color the of milk, which will happen quickly, then decant this colored vinegar, and put new distilled vinegar over the litharge, as above, so it too tints up a white color, like milk.

Decant it, and repeat this work, with distilled vinegar, until it no longer tints, and then put these colored vinegars in a glazed earthen basin, let it rest so the milky substance of Saturn goes to the bottom. Distill off the clear vinegar; the milky substance left, the nobler part, is the spirit of Saturn. It serves for many things in the enamels, and glasses. If this white material does not all go to the bottom throw fresh water over it, which will force it to the bottom. Recover what still does not precipitate by evaporating the vinegar and water, so in the bottom will be the finest part of the [milky] vinegar, good for many things in the art of glassmaking.

~ 124 ~
Rosichiero to Enamel Gold

Have crystal frit made in the following manner: Take 10 lbs of polverino salt, and 8 lbs of white tarso, ground most finely like flour. Make a paste of this material with water, so that it is a hard paste, and make it into small flat cakes. Put these in an earthen pan set into a small furnace made in the form of a limekiln, and calcine them with a good fire for 10 hours. Then move them to the neglected area of the furnace, near the eye, for 3 or 4 days until they fully calcine.

You should have calx of lead and tin, that is to say its finer parts. You should separate it with water, as demonstrated in the sixth book about enamels, in the chapter [93] about enamel base material. Take 2 lbs of this purified calx, with 2 lbs of calcined white wine tartar, mix everything thoroughly, and put it in a crucible glazed with white glass. Leave it to fuse and clarify well.

When it is ready, throw it into water, and then return it to the crucible to clarify. Throw in water a second time, reheat, and when it has clarified well add 10 oz of red copper to the crucible. Leave it to digest the color well, and then give it iron crocus made with aqua fortis [chapter 18]. Add this crocus little by little, just like manganese, and then leave it to rest for 6 hours, and watch to see if the color is good. If not then give it more of the crocus, little by little, enough to give the desired color.

~ 125 ~
Rosichiero for Gold by Another Method

Take crystal frit, made as described in the above recipe for rosichiero [chapter 124]. Fuse 4 lbs of this frit in a clean glazed crucible. When it has clarified, throw it in water. Heat it again, leave it to clarify, and throw it in water a second time. Then reheat, and leave it to clarify.

When it is well clarified, add the purified calx of lead and tin, described in the above recipe for rosichiero, add this calx little by little, ½ oz at a time, let the calx incorporate, and watch for when the glass becomes an ash gray color, at which point it will be good. Do not add too much calx because if you overload it with calx it will become white in color, which is not good.

When it turns the said gray color, do not add more calx, but leave it to clarify. Then have 2 oz fine minium, add this to the glass, and let it incorporate well, and clarify. When it clarifies well, throw into water, return it to the crucible, and leave it for 8 hours. Have ½ oz of calcined copper, that is to say red copper, as was described in its place [chapter 22], and ½ oz of raw white [wine] tartar.

Throw these things in, and stir them well. Now add a dram of hematite, which the sword makers use for burnishing, and 1 dram of fixed sulfur [chapter 126]. Stir and incorporate these powders, and watch. If it is over-colored, give it a little manganese to dilute it. If it is clear of color, add more of the fixed sulfur, hematite, a little red copper, and a little white wine tartar at your discretion so it becomes the desired color.

~ 126 ~
To Fix Sulfur for the Above Described Work

Have flowers of sulfur, and boil it in common oil for 1 hour, then remove this from the fire, and throw the strongest vinegar over it. Quickly the sulfur will go to the bottom, and the oil will float over the vinegar. Drain the oil, and the vinegar, and pour new oil over the sulfur, repeating as above. Do this again three times, and you will have fixed sulfur for the work described above [chapter 125].

~ 127 ~
A Glass as Red as Blood, That Can Serve as Rosichiero

Melt 6 lbs of lead glass, and 10 lbs of crystal frit in a crucible. When it clarifies, throw it into water, return it to the crucible, and leave it to clarify thoroughly. When it has clarified well, add calcined red copper to this glass, for example 4 or even 6 oz.

Leave it to simmer and clarify thoroughly, then add ground red wine tartar. Incorporate it with the glass, leave it to clarify, and look to see if the color appeals to you. Take a proof, and if you find it insufficiently tinted from the red copper and tartar, make it right. Take another proof, and return it to the heat [strike it], enough so that it becomes very red, this will intensify the color.

~ 128 ~
A Proven Way to Make Rosichiero

Simmer crystal frit, which has a lot of salt, as directed in the first recipe for rosichiero [chapter 124]. For example, put 6 lbs of this frit into a glazed crucible. As soon as it clarifies, add fine calx of lead and tin, which is to say the finer part, as was described in the chapter [93] for the enamel base material.

Of this calx, as an example, add 4 oz, incorporating it thoroughly in four doses. When it is well-clarified and incorporated, throw it into water, reheat, and then leave it to melt and clarify thoroughly. When it clarifies, add red copper to the glass, which we also use to make opaque red [chapters 24, 58]. For example, add 1½ oz of this. Add it in three portions, stir the glass well, and leave it to incorporate the powder, and to clarify the glass.

At the end of 2 hours, add crocus martis made with sulfur, reverberating it as discussed in its place [chapter 16]. Add 1½ oz, and stir the glass well. Add it in three portions then leave it to clarify, and to incorporate with the glass for 3 hours.

Now add 6 oz of singed tartar, with 1 oz of well-vitrified chimney soot. Singe the tartar as was done for the chalcedony [chapter 41]. With this powder mix ½ oz of crocus martis made with sulfur [chapter 16], as above. Grind these powders well, and give them to the glass in four portions. Stir it, and wait a little while from one [dose] to the next, because it will swell and make the glass boil unbelievably. When you finish adding the powder, leave the glass to clarify for 3 hours, then return to stir the glass, and make a proof in the form of a vial.

Reheat it well. If it strikes transparent red, like blood then it is good.

If it does not strike, add more singed tartar to [the glass] with soot, and crocus martis, as above. Add this material little by little until it becomes the desired color. Always let the glass rest for an hour after adding the powder, then make another vial, and reheat [strike] it. If it becomes red, like blood, and transparent then it is good, and will be very nice for enameling, as proven in Pisa many times.

~ 129 ~
Transparent Red

Calcine gold so that it becomes a red powder. This calcination is done repeatedly with aqua regis, pouring it over [the gold] five or six times. Then put this gold powder into a small earthen pan to calcine in the furnace until it becomes a red powder, which will take place after many days. Sprinkle this red powder of gold [gold chloride] over the fused glass. Use fine crystal, thrown in water many times. This gold powder, given in proportion little by little, will make the transparent red rubino glass; but you must experiment in order to find it.

~ 130 ~
The Way to Fix Sulfur for Rosichiero to Enamel Gold

Make a strong head [of foam] from [a solution of] lime and strong oak firewood ash. In this lime, boil sulfur vigorously. This leaves the lye a definite unctuous color, and combustible, due to the sulfur changing in the lye. The sulfur becomes white, and incombustible. Now thoroughly fixed, use it to make rosichiero for the goldsmiths to enamel over gold.

~ 131 ~
Vitriol of Venus, Continued From the End of Chapter 31

Take the crucibles [from chapter 31], and lute them [closed]. Place them in an open wind-furnace covered over well with burning coals. Leave them alone for 2 hours, and finally let the furnace cool. Remove the crucibles, and take out the copper, which you will find calcined to a blackish color that will exhibit a dark purple. Now calcined, grind this copper very well, and sift it through a sieve.

Have a round shaped terracotta vessel, with a flat bottom that will resist the fire. In Tuscany, we call this a tegame. I place the tegame over an opened wind-furnace on iron bars across the top, with the tegame awash in coals. Have them packed around the tegame. Put in the above said calcined copper, but first mix it with 6 oz of pulverized common sulfur for every pound of its weight.

As the heat begins to warm the tegame, and the sulfur begins to ignite and burn, you should have a long iron with a hook at the end. Mix and turn the copper continuously, so that it does not stick to the tegame. Roll it into balls, and continue this until the sulfur burns completely and there is no more smoke. Lift the heated tegame from the fire.

Take out all the copper with an iron spatula or similar tool. Grind it thoroughly in a bronze mortar, and sift it through a sieve. It will then be a black powder. Again, mix 6 oz of pulverized sulfur to every pound of copper as above, and return it to the tegame, over the furnace to set down on the iron bars. Mix the copper and sulfur together and calcine it again.

As the sulfur begins to smoke, agitate it continuously. Use the iron to scrape the copper from within the tegame. Ball it together so that it does not stick, in which cases the copper will not calcine well. Therefore, use great diligence in this important matter. When it is hot and the sulfur all stops smoking, remove it from the tegame and again grind and sift it thoroughly, which will result in a black powder.

Mix it with the weight of sulfur described above, and return it to the tegame to calcine a third time, as described. Take care at the end of this third calcination to leave the tegame alone over the fire. The copper that is inside should take on a reddish tawny color. When it has turned this color remove it from the fire, and grind it in the mortar, as above. It will be a reddish tawny powder, calcined to the level of strength known as vitriol, as promised.

~ 132 ~
Vitriol of Copper, Also Known as Vitriol of Venus, Made without Corrosives, from which is taken a Truly Vivid Light Blue, a Marvelous Thing

Now, to extract the vitriol of the above calcined copper [chapters 31, 131], have one or more large glass urinals, sized according to the amount of calcined copper. As an example put 1 lb of such copper, calcined and prepared as above, into a urinal that can hold 6 lbs of water.

Put this common clean water into the urinal with the calcined copper, and then in sand in a furnace. Give it a moderated fire for 4 hours until approximately 2 of the 6 lbs of water have evaporated, which is roughly estimate by eye. Let the furnace cool, and gently decant the water from glazed earthen pans. Put the copper that remains in the bottom into a tegame, over the furnace to evaporate all the moisture.

The water that is decanted from the pans will be tinted a deep blue color, and a beautiful wonderment. Let it stand in the pans for 2 days to settle down, so that the red part of the copper will collect in the bottom. At this point, filter this water with the usual felt cones, into a glass vessel. The part of the copper that will be in bottoms of the pans is put into the tegame, as above, to evaporate all the moisture.

When it dries, mix each pound of this copper with 6 oz of calcined sulfur. Then return it to calcine as above. Agitate it with the iron [rod] while the sulfur fumes the copper, as was described previously. To that end calcine it well, and do not permit it to stick to the tegame. Dried well, take it still warm from the tegame and grind it thoroughly, to obtain a black powder.

Mix each pound well with 6 oz of sulfur, ground as above. Return it to calcine again. For all calcinations, you must continuously churn and stir the copper with the iron. At the end of this second calcination, leave it over the fire for a long time so the copper becomes a tawny red color. When it is such, remove it from the fire and scrape it, still warm, from the tegame.

Grind it thoroughly in a bronze mortar, and sift it through a sieve, as was previously shown. Put it into a glass urinal, with 6 lbs of clean well water and set it in sand in the furnace, over a slow fire. After about 4 hours, you should see approximately 2 of the 6 lbs of water evaporated. Estimate this by eye. Then gently decant off the water, which will be tinted a most beautiful blue color.

As above, decant this in glazed pans, and let it stand for 2 days to settle down. Then filter it with the usual felt cones into a glass vessel. You will have a resplendently tinted liquid. The remaining part of the copper will be in the bottom of the pan. Take it, along with the residual copper remaining in the glass urinal, and put them into the earthen tegame. Evaporate all moisture over the furnace, as was done the other times. Give due consideration to the fact that in this work a tegame will often break.

Therefore, each time one is broken you must take a new one. Not only when it is broken, but even when it is cracked, so that it will not break over the furnace, making the copper fall into the ash and coals, causing everything to be lost. So evaporate the moisture from the copper, mix it with the usual 6 oz of pulverized sulfur to each pound of copper, and in the tegame return it over the furnace to calcine.

Always churn it with the iron [hook], as usual. At the end when the fumes stop, let it stand over the fire for a little while, until it starts to take on the red tawny color. When it is well tinted, take it warm from the tegame, grind it in a bronze mortar, and then sift it. Take the copper while it is warm because it will come out of the tegame more easily; if left to cool it sticks to the tegame, so that it is not possible to get it out.

Furthermore, the tegame is hardly ever broken if [the copper] can be detached. As has been said, always remove it warm. As usual, put the copper, ground as above, in a urinal with the usual 6 lbs of well water per pound of copper. Evaporate 2 [out of 6] lbs in the furnace, over a slow fire. Let the water cool in pans and decant as usual, letting it settle for 2 days. As above, filter as usual and it will be tinted and beautiful.

Return the above new copper to evaporate and calcine. Extract its dye into urinals with common water, as above. Filter it as before, and repeat this manipulation, not only a fourth, but fifth, and sixth time, in all and for all as above. The remaining copper is like a soft earth. The best, and most noble of its dye will all be in the filtered water, described above. Mix them all together, and filter with the usual cones of felt, for the last time. Throw the residue and filth away as useless. You will then have a liquid most limpid, and tinted a most wonderful blue color.

~ 133 ~
The Way to Extract the Vitriol From the Above Colored Water

Have a large glass urinal, capable of holding 3 flasks of liqueur. Put it in ash or sand in a furnace. Fill the urinal full of the above [chapter 132] colored water, give it a moderated fire, and start to evaporate the water. Keep other glass urinals full of the tinted water close to the furnace, in order to warm them well. From time to time, using a glass ladle, refill the large urinal in the sand, in order to evaporate [more liquid].

Keep the tinted water warm, because if you put it in cold, it will cause the larger urinal to break, and ruin everything. Keep the [other] urinals near the furnace, full of tinted water, in order to keep them warm. If for example you have 10 flasks of colored water, evaporate enough to turn it into 2½ - 3 flasks. Now the water is charged, and saturated with tincture.

Put it into glazed earthen pans,

and leave it in a cold damp place overnight. You will find the vitriol of copper has formed into crystalline points that mimic true oriental emeralds. Thoroughly decant off all the liquid in the pans, and leave the vitriol [crystals] to dry - without allowing them to stick to the pans. Evaporate half of the water, which will produce new vitriol [crystals], like those above. Repeat this until you have all the vitriol, which you should put in a retort thoroughly coated with strong lute. Take care to put no more than 1 lb of vitriol into the retort, which must not be very large, however it is good to have an amply large receiver.

Start by giving it a continuous 4 hours of a most temperate fire, because if you increase the fire even slightly then the vapors, steam, and pressure that evolve from this vitriol in the beginning are so strong, and come with such a sudden and powerful start, that no receiver could withstand it. Therefore, be sure above all else, that in the beginning you moderate the fire well for 4 hours and the joints are optimally luted.

At the end, give it a powerful fire. The spirits will start to come and dry into a white form, continue the fire until the receiver begins to clear. Let the fire die, and leave everything to cool for 24 hours. Un-lute the joints, and preserve the liqueur that is in the receiver in tightly sealed glass vessels. This is the true flaming azure blue [tincture], with which marvelous things are made. It is most potent, and as sharp as anything known in nature today, as can easily be perceived from its odor.

Many things could be said here, which are omitted as not being pertinent to the art of glassmaking, which perhaps upon another occasion you will be able to judge. A black colored residue will remain in the bottom of the retort. If you leave this for some days to the open air, by itself it will take on a lighter color. Pulverize this, and mix it with zaffer, as described above [chapter 31]. Add it to the crystal [powder] with the dose dictated. It will make a wonderful aquamarine. Although I have placed here the way to make this powder with much clarity, do not presuppose that I have described a way to make something ordinary, but rather a true treasure of nature, and this for the delight of kind and curious spirits.

THE END.

LIBRO SETTIMO

l'aria, per se sole pigliano il colore sbiadato, che sopra si dice, questo si polverizzi, & mescoli con zaffera, come sopra dandola al Cristallo, con la dose detta si farà l'acqua marina marauigliosa, & però ho posto io qui il modo di fare questa poluere con molta chiarezza presuponendomi non hauere messo vn modo di far ordinario; Ma vn vero tesoro di Natura, & questo per gusto delli spiriti Gentili, & curiosi.

IL FINE.

TABLE OF CONTENTS

The Art of Glass

FIRST BOOK

A New and Secret Method to Extract the Salt of Polverino, Rocchetta, and Soda, With Which a Crystal Frit Called Bollito, Fundamental to the Art of Glassmaking, is Made. Chap. j.   page 1

A Way to Make Crystal Frit, Otherwise Called Bollito. Chap. ij.   p.4

Another Method to Extract the Salt from Polverino, Which Makes a Crystal as Beautiful and Clear as Rock Crystal. This New Method is My Own Invention. Chap. iij.   p.6

A Caution About Golden Yellow in Crystal. Chap. iiij.   p.7

> TAVOLA DE' CAPITOLI dell'Arte Vetraria.
> LIBRO PRIMO.
>
> A Cauare il sale del Poluerino, Rocchetta è Soda con il quale si fa la fritta del Cristallo, detto Bollito, fondamento dell arte Vetraria, con vn nuouo, e secreto modo. cap.j.   a carte, 1
> Modo di fare la fritta di cristallo altrimenti detto bollire. cap. ij.   4
> Altro modo di couare il sale del poluerino, che fa il Cristallo tanto bello, e chiaro, quanto il Cristallo di montagna, modo nuouo da questo Autore inuentato. Cap.iij.   6
> Auuertimento per il giallo doro in Cristallo. cap.iiij.   7
> Modo di fare il Sale dell'Erba detta Felcie, che fa il Cristallo assai bello. cap.v.   7
> Modo di fare vn'altro sale, che farà vn Cristallo marauiglioso, & stupendo. cap.vj.   8
> Sale, che farà vn Cristallo assai bello. cap. vij.   9
> Modo di fare la fritta ordinaria, cioè di Poluerino, di Rocchetta, & di soda di Spagna. cap viij.   9
> A fare il Cristallo in tutta perfettione. cap ix.   11
> A fare il Cristallino, & vetro bianco, detto altrimenti vetro comune. cap.x.   13
> A fare il Sale di Tartaro purificato. cap.xj.   14
> A preparare la zaffera, che serue per più colori nell'Arte Vetraria. cap.xij.   15
> A preparare il Manganese per colorire i vetri. cap.xiij.   15
> A fare il ferretto di Spagna, che serue ne i colori dei vetri. c. xiiij.   16
> Altro modo di fare il detto Ferretto. cap. xv.   16
> A fare il Croco di ferro, altrimenti detto di Marte, per i colori del vetro. cap. xvj.   17
> A fare il Croco di Marte in altra maniera. cap. xvij.   17
> Altro modo di fare il Croco di Marte. cap xviij.   18
> A fare il Croco di Marte in altra maniera. cap. xix.   18
> A calcinare l'Orpello detto tremolante, che in vetro fa il coleste, & di Gazzera marina. cap. xx.   19
> A calcinare il medesimo Canterello in altra maniera, per fare il rosso trasparente, il giallo, & il Calcidonio. cap. xxj.   20
> P 2   Acqua

A Method to Make Salt from a Plant Known as the Fern, Which Makes a Very Nice Crystal. Chap. v.   p.7

A Method to Make Another Salt that will Make a Marvelous and Wonderful Crystal. Chap. vj.   p.8

A Salt that will Make a Very Beautiful Crystal. Chap. vij.   p.9

A Way to Make Ordinary Frit with Polverino, Rocchetta, and Spanish Barilla. Chap. viij.   p.9

To Make a Fully Perfect Crystal.   Chap ix.   p.11

To Make Crystallino and White Glass, Also Known as Common Glass. Chap. x.   p.13

To Make Purified Tartar Salt. Chap. xj.   p.14

To Prepare Zaffer, Which Serves for Many Colors in the Art of Glassmaking. Chap. xij.   p.15

To Prepare Manganese for Coloring Glass. Chap. xiij.   p.15

To Make Spanish Ferretto, Which Serves to Color Glass. Chap. xiiij.   p.16

Another Way to Make Spanish Ferretto. Chap. xv.   p.16

To Make Iron Crocus, Also Known as Crocus Martis, to Color Glass. Chap. xvj.   p.17

To Make Crocus Martis in Another Manner. Chap. xvij.   p.17

Another Way to Make Crocus Martis.   Chap. xviij.   p.18

To Make Crocus Martis in Another Manner. Chap. xix.   p.18

To Calcine an Orpiment Called Tinsel, Which in Glass Makes the Celestial Color of the Blue Magpie. Chap. xx.   p.19

To Calcine the Same Foils Another Way in Order to Make Transparent Red, Yellow, and Chalcedony. Chap. xxj.   p.20

Aquamarine in Glass, a Principal Color in the Art. Chap. xxÿ.   p.20

A Celestial, or Rather Blue Magpie Color. Chap xxiÿ.   p.22

A Red Copper Scale that Makes Many Colors in Glass. Chap. xxiiÿ. p.22

Thrice Baked Copper Scale for Glass Colors. Chap. xxv. p.22

Aquamarine in Artificial Crystal, Also Called Bollito. Chap. xxvi. p.23

General Warnings for All the Colors. Chap. xxvÿ. p.24

To Make Thrice Baked Copper Scale with More Ease and Less Cost then Before. Chap. xxviÿ. p.25

A Fine Crystal in Aquamarine with the Above Copper Powder. Chap. xxix. p.25

A Lower Cost Aquamarine. Chap. xxx. p.26

A Marvelous Aquamarine Beyond All Aquamarines, of My Invention. Chap. xxxi. and cxxxi. p.27, 109

Emerald Green in Glass. Chap. xxxÿ. p.27

A Nicer Green Than Above. Chap. xxxiÿ. p.28

A Marvelous Green. Chap. xxxiiÿ. p.29

Another Green, Which 'Carries the Palm' for All Other Greens, Made by Me. Chap. xxxv. p.30

Sky Blue, or More Properly Turquoise, a Principal Color

Tauola.

Acqua marina in vetro, colore principale nell'arte cap. xxÿ.	20
Colore celeste, o vero di Gazzera marina. cap. xxiÿ.	22
Ramina roffa, che serue a più colori in vetro. cap.xxiiÿ.	22
Ramina di tre cotte per i colori in vetro. cap.xxv.	22
Acqua mariana in Criftallo, artifiziale altrimenti detto bollito. cap. xxvi.	23
Auuertimenti generali in tutti i colori. cap. xxvÿ.	24
A fare Ramina di tre cotte con più facilità, & manco spesa della sopradetta. cap. xxviÿ.	25
Acqua marina in cristallo bella, con la sopradetta Ramina. cap. xxix.	25
Acqua marina di manco spesa. cap.xxx.	26
Acqua marina marauigliosa sopra tutte l'acque marine di mia inuentione. cap. xxxi. & cxxxi.	à 27. è 109
Verde smeraldino in vetro. cap. xxxÿ.	a 27
Uerde più bello del sopradetto. cap. xxxiÿ.	28
Uerde marauigliofo. cap. xxxiiÿ.	29
Altro verde che porta la palma di tutti gli altri. verdi per me fatti. cap. xxxv.	30
Aierino, o vero di Turchina, colore principale nell'Arte Vetraria. cap.xxxvi.	30

LIBRO SECONDO

MODO di calcinare il tartaro, & vnirlo con il Rofichiero, che fa apparire i vaghi scherzi di molti colori con ondeggiamenti in efsi, & gli da l'Opaco, come anno de naturali orientali. cap. xxxvÿ.	33
Uodo di fare l'acqua forte, detta da partire, che folue l'Argento & l'Argento viuo, con un modo fegreto. cap.xxxviÿ.	35
A purificare il vetriolo per fare vn'Acqua forte potentiffima. cap. xxxix.	38
A fare l'Acqua Regia che folue l'Oro, & li altri metalli dall'Argento in fuora cap. xl.	38
Abruciare il Tartaro detto Greppola di vino cap. xli.	39
A fare il Calcidonio in vetro affai bello cap. xlÿ	39
Secondo Calcidonio Cap. xliÿ	41
Terzo modo di Calcidonÿ cap. xliiÿ.	44

in the Art of Glassmaking. Chap. xxxvi.   p.30

SECOND BOOK

The Way to Calcine Tartar Will be Covered, as well as How to Unite it with Rosichiero. This Causes the Lovely Play of Many Undulating Colors to Appear, with the Same Shading as the Natural Oriental Stones. Chap. xxxvÿ.
    p.33

How to Make Nitric Acid, also Called Parting Water, which Dissolves Silver and Mercury, by a Secret Method. Chap. xxxviÿ.   p.35

To Purify Vitriol in Order to Make a Most Potent Acid. Chap. xxxix.   p.38

To Make Aqua Regis, which Dissolves Gold, and Other Metals from Silver. Chap. xl.   p.38

To Sear the Tartar of Wine Dregs. Chap. xli.   p.39

To Make a Very Beautiful Chalcedony in Glass. Chap. xlÿ   p.39

A Second Chalcedony. Chap. xliÿ   p.41

A Third Way to Make Chalcedony. Chap. xliiÿ.   p.44

THIRD BOOK

To Make Golden Yellow in Glass. Chap. xlvi.   p.50

Garnet Color. Chap. xlvij.   p.51

Amethyst Color. Chap. xlviÿ.   p.51

Sapphire Color. Chap. xlix. p.52

A Prettier Sapphire Color. Chap. l. p.52

A Black Color. Chap. li. p.53

A Prettier Black Color. Chap. lÿ. p.53

An Even Prettier Black. Chap. liÿ. p.53

A Beautiful Lattimo. Chap. liiÿ. p.53

A Beautiful and Whiter Lattimo. Chap. lv. p.54

To Make Marble Color. Chap. lvi. p.54

Peach Blossom in Lattimo. Chap. lvÿ. p.54

Opaque Red. Chap. lviÿ. p.55

Frit of Rock Crystal. Chap. lix. p.56

Pearl Color in Crystal. Chap. lx. p.56

FOURTH BOOK

To Calcine Lead. Chap. lxÿ. p.58

To Make Lead Glass. Chap. lxiÿ. p.59

How to Work Lead Glass. Chap. lxiiÿ. p.59

Tauola.

LIBRO TERZO.

A Fare il Giallo Doro in vetro cap.xlvi.	50
Colore di Granato cap.xlvij.	51
Colore di Amatisto cap.xLviÿ.	51
Colore di zaffiro cap.xLx.	52
Colore di zaffiro più bello cap.L.	52
Colore Nero cap.Li.	53
Colore Nero più bello. cap.lÿ.	53
Altro nero più bello.cap.liÿ.	53
Lattimo bello.cap.liiÿ.	53
Lattimo bello,e più bianco. cap.lv.	54
A fare marmorino.cap.lvi.	54
Perfeghino in lattimo.cap.lvÿ.	54
Rosso in corpo.cap.lviÿ.	55
Fritta di cristallo di montagna. cap.lix.	56
Colore di Perla in Cristallo. cap.lx.	56

LIBRO QVARTO.

A Calcinare il piombo.cap. lxÿ.	58
A far il vetro di piombo. cap.lxiÿ.	59
Modo di lauorare detto vetro cap. lxiiÿ.	59
Uetro di piombo in colore smeraldino mararngliofo. cap.lxv.	60
Altro verde smeraldino marauigliofo sopra tutti i verdi. c.lxvi.	62
Colore di Topatio in vetro piombo. cap.lxvÿ.	62
Colore celeste,o vero di Gazera marina in vetro di piõbo. c.lxviÿ.	62
Colore di Granato nel vetro di piombo. cap.lxix.	63
Colore di zaffiro,in vetro di piombo. cap. lxx.	64
Colore di giallo d'Oro in vetro di piombo. cap. lxxi.	64
Colore di Lapis Lazzuli.cap. lxxÿ	65
Modo di tingere il Cristallo di Montagna senza fondere in Colore di vipera cap. lxxÿ.	65
Colore di Balascio Rubino, Topatio, Opale, & Girasole, nell'Cristallo di montagna, cap. lxxiv.	66

LI-

Lead Glass in A Marvelous Emerald Green Color. Chap. lxv. p.60

Another Marvelous Emerald Green Beyond All the Greens. Chap. lxvi. p.62

Topaz Color in Lead Glass. Chap. lxvÿ. p.62

A Celestial, or True Azure Magpie Color in Lead Glass. Chap. lxviÿ. p.62

Garnet Color in Lead Glass. Chap. lxix. p.63

Sapphire Color in Lead Glass. Chap. lxx. p.64

The Color of Golden Yellow in Lead Glass. Chap. lxxi. p.64

The Color of Lapis Lazuli. Chap. lxxÿ. p.65

A Method to Tint Rock Crystal the Color of a Viper Without it Melting. Chap. lxxiÿ. p.65

The Colors Balas, Ruby, Topaz, Opal, and Girasol in Rock Crystal. Chap. lxxiv. p.66

FIFTH BOOK

The Way to Prepare Rock Crystal. Chap. lxxvj. p.70

The Way to Make Oriental Emerald. Chap. lxxvij. p.71

An Emerald More Charged with Color. Chap. lxxviij. p.73

To Make a More Graceful Emerald Paste. Chap. lxxix. p.74

Another Most Attractive Emerald. Chap. lxxx. p.74

Oriental Topaz. Chap. lxxxj.   p.75

Oriental Chrysolite. Chap. lxxxij.   p.75

A Celestial Color. Chap. lxxxiij.   p.75

A Celestial Color with Violet. Chap. lxxxiiij.   p.76

Oriental Sapphire. Chap. lxxxv.   p.76

An Oriental Sapphire Loaded with Color. Chap. lxxxvj.   p.76

Oriental Garnet. Chap. lxxxvij.   p.77

A Deeper Oriental Garnet. Chap. lxxxviij.   p.77

Another Beautiful Garnet. Chap. lxxxix.   p.78

Advice about Pastes, and their Colors. Chap. xc.   p.78

A Marvelous Way, No Longer Used, to Make the Above Pastes, and to Imitate Every Type of Gem. Chap. xcj.   p.79

A Way to Make Hard Pastes in All the Colors. Chap. xcij. p.81

SIXTH BOOK

Materials With Which to Make All Enamels. Chap. xciij.   p.84

Tauola.
LIBRO QVINTO.

Modo di preparare il Cristallo di montagna cap. lxxvj.   70
Modo di fare lo smeraldo Orientale cap. lxxvij.   71
Smeraldi più carichi di Colore. cap. lxxviij.   73
A fare Pasta di smeraldi più vaga cap. lxxix.   74
A fare smeraldo bellissimo cap. lxxx.   74
Topatio Orientale, cap. lxxxj.   75
Grisopatio Orientale cap. lxxxij.   75
Colore Celeste cap. lxxxiij.   75
Colore Celeste col violino. cap. lxxxiiij.   76
Zaffiro Orientale cap. lxxxv.   76
Zaffiro Orientale Carico di Colore cap. lxxxvj.   76
Ingranato Orientale cap. lxxxvij.   77
Ingranato Orientale più carico cap. lxxxviij.   77
Altro Ingranato bello cap. lxxxix.   78
Auuertimenti per le paste e loro colori. cap. xc.   78
Modo di fare le sopradette paste, & imitare ogni sorte di gioie, marauiglioso, & non più vsato. cap xcj.   79
Modo di fare le paste di tutti i colori durissime. cap. xcij.   81

LIBRO SESTO.

Materia con laquale si fanno tutti li smalti. cap. xciij.   84
Smalto bianco lattato. cap. xciiij.   84
Smalto turchino. cap. xcv.   85
Altro smalto azzurro. cap. xcvj.   86
Smalto verde. cap. xcvij.   87
Altro smalto verde. cap. xcviij.   87
Altro smalto verde. cap. xcix.   88
Smalto nero. cap. c.   88
Altro smalto nero. cap. cj.   88
Altro smalto nero. cap. cij.   89
Smalto auuinato cap. ciij.   89
Smalto pauonazzo. cap. ciiij.   90
Smalto giallo. cap. cv.   90
Smalto celeste. cap. cvj.   90
Smalto violato. cap. cvij.   91

LI-

Milk White Enamel. Chap. xciiij.   p.84

[Two] Turquoise Enamels. Chap. xcv.   p.85

Another Blue Enamel. Chap. xcvj.   p.86

Green Enamel. Chap. xcvij.   p.87

Another Green Enamel. Chap. xcviij.   p.87

Another Green Enamel. Chap. xcix.   p.88

Black Enamel. Chap. c.   p.88

Another Black Enamel. Chap. cj.   p.88

Another Black Enamel. Chap. cij.   p.89

Red Wine Colored Enamel. Chap. ciij.   p.89

Purple Enamel. Chap. ciiij.   p.90

Yellow Enamel. Chap. cv.   p.90

Celestial Blue Enamel. Chap. cvj.   p.90

Violet Enamel. Chap. cvij.   p.91

SEVENTH AND FINAL BOOK

A Yellow Lake From Broom flowers, for Paint. Chap. cviij.   p.94

A way to Extract Lake from Poppies, Blue Irises, Red Roses, Violet Roses, and From All Kinds of Green Plants. Chap. cix.   p.95

A Way to Extract the Lake and Color for Painting, From Orange Blossoms, Red Poppies, Blue Irises, Ordinary Violets, Red Violets, Carnations, Red Roses, Borage Flowers, Day Lilies, Irises, and From Flowers of Any Desired Color, and the Greens of the Mallow, the Pimpernel, and All the Plants. Chap. cx.   p.95

An Azure [Paint] like Alemagna Blue. Chap. cxj.   p.96

The Way To Color Faded Natural Turquoise. Chap. cxij   p.96

A Mixture to Make [Mirror] Spheres. Chap. cxiij.   p.96

The Way to Tint Glass Balls, and Others Vessels of White Glass, From the Inside, in All Kinds of Colors, so That They Will Imitate Natural Stones. Chap. cxiiij.   p.97

Ultramarine Blue. Chap. cxv. p.98

[Prepared Wool Shearings to Make a] Lake of Kermes for Painters. Chap. cxvj.   p.99

> ## Tauola.
> ### LIBRO SETTIMO, ET VLTIMO.
> Lacca gialla per dipignere, da i fiori di Ginestra. cap. cviij.   **94**
> A cauare la lacca di Rofolacci, Fioralifi, Rofe roffe, Viole roffe, & da ogni forte di erba verde. cap. cix.   **95**
> A cauare la lacca, & colore per dipingere da fior Ranci, Rofolacci, Fioralifi, Viole ordinarie, Viole roffe, Rofe incarnate, Rofe roffe, Fiori di Borrana, Fiori di Capucci, Fiori di Ghiaggiuolo, & da ogni fiore di qual fi voglia colore, & il verde della Malua, della Pimpinella, & di tutte l'erbe. cap. cx.   **95**
> Azzurro come quello di Alemagna. cap. cxj.   **96**
> Modo di colorire le Turchine naturali fcolorite. cap. cxij   **96**
> Meftura da fare le Spere. cap. cxiij.   **96**
> Modo di tingere palle di vetro, ò altri vafi di vetro bianco, per di dentro d'ogni forte di colori, che imiteranno le pietre naturali. cap. cxiiij.   **97**
> Azzurro oltramarino. cap. cxv.   **98**
> Lacca di Chermesì per Pittori. cap. cxvj.   **99**
> Maeftra per cauare il colore del Chermefi. cap. cxvij.   **100**
> Lacca del verzino è della Robbia affai bello. cap. cxviij.   **102**
> Lacca di Chermifi in altra maniera è più facile. cap. cxix.   **103**
> Roffo trafparente in vetro. cap. cxx.   **103**
> Roßo come fangue. cap. cxxj.   **104**
> Colore di Balafcio. cap. cxxij.   **104**
> A cauare l'anima di Saturno, che ferue a molte cofe nelli fmalti e vetri. cap. cxxiij.   **105**
> Rofichiero per fmaltar loro. cap. cxxiiij.   **105**
> Rofichiero d'oro in altra maniera. cap. cxxv.   **106**
> A fiffare zolfo per l'opera foprafcritta. cap. cxxvj.   **107**
> Vetro roßo, come fangue, che puol feruire per Rofichiero. cap. cxxvij.   **107**
> Modo di fare il Roficchiero prouato. cap. cxx viij.   **107**
> Roffo trafparente. cap. cxxix.   **108**
> Modo di fiffare il zolfo per il Rofichiero da fmaltare oro. cap. cxxx.   **109**
> Vetriolo di Uenere, che comincia in quefto, nella fine del capitolo. 31. a 27. cap. cxxxj.   **109**
> Vetriolo di rame, altrimenti detto di Venere fenza corrofini, del quale fi caua il vero acceso azzurino, cofa marauigliofa. c. cxxxij. **110**
> Modo di cauare il vitriolo da dette acque colorite. cap. cxxxiij   **112**
> *IL FINE.*

An Elixir for Extracting the Color From Kermes. Chap. cxvij. p.100

Very Beautiful Lakes from Brazil Wood, and Madder Root. Chap. cxviij.   p.102

Lake of Kermes Another Easier Way. Chap. cxix.   p.103

Transparent Red in Glass. Chap. cxx.   p.103

Red Like Blood. Chap. cxxj.	p.104
Balas Color. Chap. cxxij.	p.104
To extract the Spirit of Saturn, Which Serves Many Uses in Enamels and Glasses. Chap. cxxiij.	p.105
Rosichiero to Enamel Gold. Chap. cxxiiij.	p.105
Rosichiero for Gold by Another Method. Chap. cxxv.	p.106
How to Fix Sulfur for the Above Described Work. Chap. cxxvj.	p.107
Blood Red Glass, Which Can Serve for Rosichiero. Chap. cxxvij.	p.107
A Proven Way to Make Rosichiero. Chap. cxxviij.	p.107
Transparent Red. Chap. cxxix.	p.108
The Way to Fix Sulfur for Rosichiero to Enamel Gold. Chap. cxxx.	p.109
Vitriol of Venus, Continued From the End of Chapter 31. Chap. cxxxj.	p.109
Vitriol of Copper, Also Known As Vitriol of Venus, Made Without Corrosives, From Which is Taken the Truly Vivid Light Blue, a Marvelous Thing. Chap. cxxxij.	p.110
The Way to Extract Vitriol From the Above Colored Water. Chap. cxxxiij.	p.112

THE END

COPY OF PERMISSION BY THE MOST HOLY SUPERIORS

The M.R. Mr. Filippo del Migliore of the Florentine Archdiocese reviewed the present work, considering if in it there is anything that proceeds counter to Christian conscience, and good customs, and testifies to this below.

Piero Niccolini, Vicker of Florence 30. March.1612

I Filippo del Migliore of the Florentine Archdiocese have reviewed the present work, and have not found in it anything that proceeds counter to Christian conscience, and good customs, and have testified so in writing and signed below in my own hand this 2nd, of April 1612.

Filippo del Migliore of the Archdiocese.

Pending the preliminary report we grant, that the above written work can be printed in Florence, the usual orders being issued.

Piero Nicolini Vicker. of Florence. 2. April 1612.

The P.M. Agostino Vigiani, Regent of Servants, reviewed the present work for part of the Holy Office and testifies. From the Holy Office.

In Florence, 2. April 1612.
F. Corn, Inquisitor of Florence.

I have read the present Work of the Art of Glassmaking, in which I have not found anything repugnant to the Christian conscience, and good customs, but full of things, and natural secrets, of no small use, and curiosity.

4. April 1612.
Me, F. Agostino Vigiani, Regent of Servants (Signed in my own hand.)

F. Corn. Inquisitor of Florence 4 April 1612.

Niccolò dell'Antella; Imprint in accordance with these orders of 7 April 1612.

Registered.

+ ABCDEFGH IKLMNOP.
All are full sheets.

GLOSSARY

Appearing below are specialized terms uses throughout the book. The terms used by Neri appear both in the original renaissance Italian, and in their equivalent modern English translations. For these words, the English entries contain references leading to the original Italian terms, where the appropriate descriptions are found.

Aceto - *ital.* vinegar, the active ingredient of which is acetic acid (CH_3COOH) also known as ethanoic acid. This weak acid is responsible for vinegar's characteristic sour taste, acid flavor, and pungent odor. It forms through the oxidation of ethyl alcohol, and in its pure form is a colorless liquid. Used by Neri primarily to accelerate the oxidation of metals which are used for the pigmentation of glass.

Acciaio - *ital.* a term currently used for the metal steel (Fe+C), however in the early seventeenth century steel was called 'ferro duro' (hard iron), and 'acciaio' was a synonym for bronze (Cu+Sn). Today, the Romanian term 'acioaie' is still used for bronze, derived from the archaic Italian. Neri uses it in chapter 113 cast into the form of spherical mirrors, in the proportion of 75% tin, 25% copper, with a pinch of arsenic. This mixture results in a white metal; as tin content passes 20% the alloy rapidly loses it characteristic yellow-brown color, hence 'Bronzo' for the low tin alloy of copper, and 'acciaio' for the high tin alloy. The arsenic in this recipe serves both to harden and brighten the composition, allowing a high polish. Because spherical mirrors have the properties of optical lenses, they were sometimes referred to as 'vetri acciaio' (steel glasses). Similar 'white bronze' alloys were used to make mirrors in China as early as 2000 BCE, and have also been employed very effectively to counterfeit silver. Another synonym is 'speculum' which is Latin for 'metal mirror', and refers both to the object as well as the alloy.

Acqua forte - *ital.* literally 'strong water', usually referred to nitric acid (HNO_3). Neri also uses this term to describe mixtures of nitric and sulfuric acid, and sometimes as a generic term for acid. Nitric acid was produced by heating a mixture of alum, green vitriol, and saltpeter, and collecting the evaporated HNO_3. See chapter 38-39.

Acqua partire - *ital.* literally 'parting water', Neri uses it as a synonym for nitric acid = 'aqua fortis', while other references equate it with aqua regis.

Acqua regis - *ital., engl.* literally 'king water'. A mixture of Nitric and hydrochloric acids (HNO_3 + HCl) in various proportions, depending on the material to be dissolved. Commonly, more nitric acid than hydrochloric was employed. Dubbed the 'king' of acids because it dissolves gold, the 'king' of metals.

Acqua vite - *ital.* also acquavite, acquavita; a term today synonymous with any strong distilled liquor especially brandy, rum, or schnapps. Neri was almost certainly referring to grappa, a traditional drink of Italy produced by distilling fermented grape pulp and solids left over from wine production. The result (when done correctly) is a strongly flavored beverage having a high ethyl alcohol content (Aqueous Ethanol, C_2H_5OH.) Grappa bianca (white or un-aged grappa) is crystal clear, and would have been ideal for dissolving the organic pigments in flowers and plants (chapter 110). Commercially produced grappa contains about 40% by volume of alcohol, with homemade stills producing output of even greater potency.

Agata - *ital.* agate, a form of chalcedony (quartz SiO_2) characterized by strong concentric banding of a variety of colors. From ancient times, agate was regarded as having special protective powers, and throughout the renaissance, bowls and other items made of agate were collected enthusiastically by the aristocracy. See also *calcidonio*.

Agate - *engl.* see agata.

Alchali - *ital.* alkali, a term derived from the Arabic *al* = from + *kali* = 'the kali plant'. A costal and desert shrub found around the world whose ash is very high in the metals potassium (atomic symbol K for kali) and calcium (Ca), two key ingredients of the so-called soda-lime glass produced in Neri's time. In modern chemistry the term alkali refers to a group of very reactive metals on the periodic chart, which do not freely occur in nature, but always in compounds such as salts, carbonates and oxides. These compounds are the fluxes which make glass flow at reasonable temperatures. The alkali group includes Lithium (Li), sodium (Na), potassium (K), rubidium (Rb), cesium (Cs), and francium (Fr).

Alkali salt - *engl.* see *sale alchali*.

Alemagna - *engl.* see *azzurro d'alemagna*.

Alembic - *engl.* see *cappello di vetro*.

Alum di cantina - *ital.* potash, literally 'cellar alum'. An alkaline potassium compound, especially potassium carbonate or hydroxide. It was obtained by leaching vegetable ashes in boiling water and evaporating the solution (lixiviation). This is the general case of the same process used by Neri in the first chapters of the book to obtain salt from Rocchetta and Polverino for making basic glass and crystal.

Allume di rocho - *ital.* roche alum, also called Roman alum. The mineral Alumstone, a gray or pinkish form of alunite found in volcanic rocks. In its pure form alum is a colorless to white crystal composed of potassium aluminum sulfate $KAl(SO_4)_2 \cdot 12H_2O$ or $[KAl_9H_2O)_6]SO_4 \cdot 6H_2O$. Used as a mordant in dyes, and an astringent in medicines. It precipitates out of a water solution, forming positively charged particles which attract negatively charged organic impurities thus purifying the water. Alum is also used in the production of nitric acid (aqua fortis). The name derives from Rocca, in Syria, where alum is said to have been obtained. See also *allumina*.

Allumina - *ital.* to mordant silks, wools or other fabrics with alum before they can be dyed. The mordanting process causes a permanent chemical bond to form between the fibers and the dyestuff. See also *allume di rocho*.

Alum - *engl.* see *allume di rocho*.

Alumstone - *engl.* see *allume di rocho*.

Amalgam - *engl.* see *malgama*.

Amatisto - *ital.* amethyst, a transparent violet or purple gem variety of quartz (SiO_2). Its coloration is due to small amounts of manganese and iron oxide. The deepest colored stones are highly valued as gems. From the Greek *amethustos* = 'not drunken' because the stone was believed to prevent intoxication.

Amethyst - *engl.* see *amatisto*.

Ammoniac - *engl.* see *sale ammoniaco*.

Ammonium chloride - *engl.* see *sale ammoniaco*.

Ana - *ital.* a term used by doctors and pharmacists meaning equal amounts. This is not a glassmaking term per se, and perhaps shows the influence of Neri's physician father, as well as his own keen interest in the subject of herbal remedies. In several glass recipes, he refers to ingredients as 'medicine'.

Anima di Saturno - *ital.* the spirit (or soul) of Saturn. For Neri's purposes an alchemical preparation consisting of lead acetate. Lead oxide dissolved in vinegar (dilute acetic acid) affords lead acetate, (see Zucchero di Saturno) also known as saccharum saturni, or sugar of lead. Lead acetate upon heating gives acetone (propanone) in the reaction $Pb(CH_3CO_2)_2 = PbO + CH_3COCH_3 + CO_2$. It is this impure acetone that is commonly known in alchemical texts as 'Spirit of Saturn', however Neri, in chap. 123 is clearly referring to the acetate, which he apparently did not realize was equivalent to *Zucchero di Saturno*.

Annealing chamber - *engl.* see *era*.

Antimonio crudo - *ital.* crude antimony, ore containing 90% or more of the metal antimony and the remainder typically (in Italy) lead and sulfur-antimony compounds such as the mineral kermesite (Sb_2S_2O). Antimony is a bluish-white metal, and element 51 on the periodic table. In glass Neri uses it in the formation of 'luster' finishes (chapter 74), and as in his chalcedony recipes (chapter 43).

Antimony - *engl.* see *antimonio crudo*.

Apostles' creed - *engl.* see *credo*.

Aqua fortis - *engl.* see *acqua forte*.

Aqua regia - *engl.* see *acqua regis*.

Aqua vite - *engl.* see *acqua vite*.

Arancio - *ital.* see *rancio*.

Argento - *ital.* the metal silver (Ag). This white shiny precious metal is atomic number 47 on the periodic chart. In its pure state silver is very malleable and ductile. It naturally occurs in its pure state, as a chloride (AgCl) ceragyrite and as a sulfide (Ag_2S) acanthite. The pure metal is soluble in nitric or hot sulfuric acids.

Argento vivo - *ital.* also *argentovivo*, literally 'live silver' see *mercurio*.

Arsenic - *engl.* see *arsenico christallino*.

Arsenico christallino - *ital.* crystalline arsenic, also called white arsenic. A highly poisonous naturally occurring crystalline form of arsenic trioxide (As_2O_3). The elemental metal is a steel gray color and atomic number 33 on the periodic chart. In glass arsenic is used commonly as a flux, to bring the melting temperature down, to facilitate the expulsion of gas bubbles, and in larger quantities as an opacifier. Neri uses it in the production of acids, in his chalcedony recipes, and in the tinting of rock crystal through vapor deposition. See also orpimento. Arsenic is highly poisonous in all its forms, and prolonged low-level exposure has been linked closely to bladder cancer.

Artificial (rock) crystal - *engl.* see *bollito*.

Assay - *engl.* see *gustare*.

Avinato - *ital.* also *avvinato*, the characteristic purplish color of red wine. A term also sometimes used to denote a taste, or aroma reminiscent of red wine or vinegar.

Avoirdupois - A system of weights, originally French, now adopted throughout the world. It forms the cornerstone of the 'customary system' in the United States.

Azure (winged) magpie - *engl.* see *gazzera marina*.

Azurite Blue - *engl.* see *azzurro d'alemagna*.

Azzurro d'alemagna - *ital.* also known as *azzurro della magna, azzurro di montagna, azzurro tedesco, azzurro citramarino*. A painter's pigment made from the mineral azurite $2(CuCO_3)\cdot Cu(OH)_2$. Until the 1600's it came mostly from central Europe. It is basic copper carbonate in its natural state that is extracted from copper mines mixed with malachite. Used since antiquity, Pliny called it "lapis armenius". It was often preferred to lapis-lazuli as an economical alternative. In alfresco painting it has a tendency to blacken; with humidity it transforms into green malachite.

Bacchettina - *ital.* see *bastoncino*.

Bagno maria - *ital.* bainmarie, or double-boiler. A pan of hot water in which a cooking container is placed for slow cooking, as temperature is limited to the boiling point of water. This was the first method for controlling distillation. From Latin balneum Mariae 'bath of Maria', the invention of which is attributed to the great Alexandrian alchemist Marie the Jewess around 200 CE.

Bain-marie - *engl.* see *bagno maria*.

Balas - *engl.* see *balascio*.

Balas ruby - *engl.* see *balascio*.

Balascio - *ital.* balas ruby, rubicelle, sometimes called false ruby, a spinel ($MgAl_2O_4$) of a delicate rose-red to orange variety. From Persian Badaksan, a district of Afghanistan where lapis lazuli is also mined.

Barilla - *ital.* synonym for 'soda', the Spanish term for plant ash rich in alkali (potassium) salts.

Basin - *engl.* see *catinella*.

Bastoncino - *ital.* a small stick, or as in this case Neri is referring to a stirring rod. Also called a *bacchettina*.

BCE - *engl.* abbreviation for Before Common Era, an indication used after dates, which is a secular equivalent of BC (384 BCE = 384 BC). See CE.

Beadmaking - *engl.* see *conteria*.

Beaker - *engl.* see *bicchiere*.

Berrettino - *ital.* a gray or ashen color.

Biacca - *ital.* see *cerusa*.

Biadetto - *ital.* a synonym for celestial, or ultramarine blue, diminutive of *biado*, or *biavo*. A term dating at least back to the 13th century.

Bicchiere - *ital.* a drinking glass, or beaker.

Black tar - *engl.* see *pecie nera*.

Black salt - *engl.* see *sal nero*.

Blue enamel paint - *engl.* see *smalto azzurro*.

Blue iris - *engl.* see *fioraliso*.

Blue magpie - *engl.* see *gazzera marina*.

Boccia di vetro - *ital.* literally glass body, a flask used by alchemists, with a narrow or wide neck that can be easily sealed or cemented to other glassware, and heated in the furnace. Equivalent to modern day Erlenmeyer and Florence flasks used by chemists. see also *fiaschi*.

Boccietta - *ital.* a small vessel, or perfume bottle, a vial.

Boil - *engl.* see *quocere*.

Bollito - *ital.* glass frit used specifically for higher quality crystal glasses.

Borage - *engl.* see *borrana*.

Borrana - *ital.* borage, a herbaceous plant with bright blue flowers and hairy leaves. *Borago officinalis* from the Latin *vorago*, and perhaps from Arabic *abu huras* = 'father of roughness' (referring to the leaves).

Bowl - *engl.* see *pignattino, scodella*.

Brazilwood - *engl.* see *verzino*.

Brick - *engl.* see *mattone*.

Brimstone - *engl.* see *suolo di zolfo*.

Bronze - *engl.* see *acciaio*.

Broom flower - *engl.* see *ginestra*.

Cabbage flower - *engl.* see *fior cappucci*.

Cake - *engl.* see *pastello, calcinacci*.

Calamine - *engl.* see *zelamina*.

Calcara - *ital*, an oven used to calcine (roast) materials for glassmaking. Used especially for frit production. Also known as a limekiln.

Calce - *ital*, see *calcie*.

Calcidonio - *ital*, chalcedony an umbrella term encompassing a wide variety of the so-called cryptocrystalline quartz gemstones. It is a form of quartz composed of microscopic interlocking crystals with an overall waxy translucent luster. Varieties include agate, bloodstone, carnelian, chrysoprase, jasper, onyx, sard, and sardonyx. The overall appearance is often banded and the range of colors include red, brown, orange, blue, purple, green, white, grey, and black. The colors are caused by various impurities. Chalcedony glass is generally a translucent white, grey, or reddish-brown ground with bands or swirls of other colors mixed in a random pattern throughout. It was and still is highly valued for its beauty.

Calcie - *ital*, calx, (plural: calces), powdery metallic oxide formed when an ore or mineral has been heated. Neri uses it to describe any oxidized (sic. calcined) metal. Probably from Gk khalix 'pebble, limestone'. He reserves the terms *sale de calcina*, and *calcinacci* to specifically describe lime (CaO) products.

Calcie di piombo - *ital*, lead calx, which is a white colored lead oxide (PbO) produced by heating metallic lead in a kiln. In chapter 62 Neri describes a second calcination process in which a yellow material is formed, this is probably a mixture with *minium* Pb_3O_4, another oxide state of lead that is orange colored.

Calcie di stagno - *ital*, tin calx, which is tin oxide (SnO_2). This off-white powder is produced by heating metallic tin in a kiln. Neri uses is as glass opacifier.

Calcina - *ital*, see *calcinare, calcie, sale della calcina*.

Calcinacci - *ital*, lime cake, apparently a form of processed lime (CaO), which Neri uses in chapter 38 in his recipe for nitric acid. It may have been a convenient form for shipping once it was produced by heating limestone or seashells in a furnace.

Calcinare - *ital*, to calcine, reduce, oxidize, or desiccate by roasting or strong heat. The term comes from the roasting of calcium carbonate ($CaCO_3$) in order to form lime (CaO), a process invented before the first century B.C. by the Romans to make cement. Neri uses it as a general term for roasted powder, usually a metal oxide. He reserves the terms *calcina, sale de calcina*, and *calcinacci* to specifically describe lime (CaO) products.

Calcinati - *ital*, masculine plural past participle of calcinare, used by Neri as a noun form, simply meaning 'calcined material'. Not to be confused with calcinacci see calcinare.

Calcine - *engl.* see *calcinare*.

Calcium oxide - *engl.* see *sale della calcine*.

Caldaro - *ital*, a kettle or cauldron. See also *paioletto*.

Calderaio - *ital*, a kettle-smith, pl: calderai. A craftsman skilled in the art of working copper. A tinker.

Calx - *engl.* see *calcie*.

Camoza - *ital*, chamois cloth, a soft absorbent leather made from the skin of sheep, goats, or deer. The term comes from the agile goat-antelope with short hooked horns, found in mountainous areas of southern Europe.

Cane - *engl.* see *canna*.

Canna - *ital.* a term used variously by Neri to describe both glass lampworking cane, and the iron punty (pontil) used to gather glass from the furnace pot.

Canterello - *ital.* tinsel, or so called gold foil; composed of thin copper sheet treated with calamine to form a gold colored brass.

Cappello di vetro - *ital.* literally glass head, a piece of glassware used by alchemists for distillation. Also known as an alembic, it is a gourd-shaped container, open on the bottom, which is connected to the top of a flask. At the top it has a long neck or beak leading downward for conveying the products to a receiver. From Arab. *al-'anbik*: *al* = 'the' + *anbik* = 'still'

Cappucci - *ital.* see *fior cappucci*.

Capo morto - *ital.* an alchemical term describing the solid residue left in the bottom of a receiver after a distillation process is completed. Neri uses it in connection with a 'dry distillation' process described in chapter 31.

Caput mortum - *engl.* see *capo morto*.

Caricare - *ital.* to load; Neri uses this term to describe the process of adding colorant to a glass batch. See *scaricare*.

Carnation - *engl.* see *rose incarnate*.

Catinella - *ital.* a tray, pan, or basin. See also *tegami, teglia*.

Cauldron - *engl.* see *caldaro, paioletto*.

CE - *engl.* abbreviation for Common Era, an indication used after dates, which is a secular equivalent to the use of AD (1576 CE = AD 1576). Historical dates previous to the year 1 are referred to as BCE or Before the Common Era which is equivalent to the use of BC (384 BCE = 384 BC). This simple change has been officially adopted by the British Government, and is increasingly used by scholars around the world. Its significance is that dates can be referenced without specific religious significance. AD abbreviates the Latin Anno Domini = the year of Our Lord (years after the birth of Christ), and BC= Before Christ. By referencing the start of the (Gregorian/Julian) calendar instead of the birth of Christ (which was probably not in the year 1 anyway) a more universal system results. see *BCE*.

Cera - *ital.* wax of any kind. Made from a variety of sources, both animal and vegetable, which are a combination of fatty acids with higher alcohols. Beeswax from the honeybee was the most popular, but other sources included sperm whales, and the carnauba (Brazilian fan palm).

Cerus[s]a - *ital.* also called *biacca*. White lead (lead carbonate), a white pigment with excellent covering ability discovered in antiquity, derived from the vitreous white or clear crystalline mineral cerussite ($PbCO_3$) lead carbonate, sometimes in a mixture with lead hydroxide. Extremely toxic due to its rapid absorption into the body, historically a major cause of lead poisoning among painters and fine artists.

Ceruse - *engl.* see *cerusa*.

Cerusite - *engl.* a mineral consisting of lead carbonate ($PbCO_3$), mined in the Veneto region of Italy since before the Roman Empire. See *cerusa*.

Chalcedony - *engl.* see *calcidonio*.

Chamber Pot - *engl.* see *Orinale di vetro*.

Chamois - *engl.* see *camoza*.

Chermisi - *ital.* also kermes, *cremisi* = crimson. A highly sought red dye, originally obtained from the dried bodies of pregnant females of the kermes shield-louse, *Coccus illicis*. A scaled insect found in the far-east and also around the Mediterranean. They make berry-like galls on the kermes oak, from which they were picked dried and sold as a commodity. In 1464, Pope Paul II decreed that chermisi should be used exclusively as the Cardinals' purple. In the 16th century Cortés brought a similar insect back from Mexico; the grana, or cochineal which had been used by the Aztecs. This insect also a shield-louse produced a much more potent dye, small amounts yielding an intense color, its native habitat being the prickly-pear cactus. By the mid 1500s the Spanish monopoly of cochineal had completely supplanted the use of the kermes, although the old nomenclature stuck. See also chapter 117. Interestingly, Neri uses alum as a mordant for this dyestuff; the advent of using tin oxide (SnO_2) to make these dyes truly colorfast was not discovered until later.

Chrysolite - *engl.* see *grisopatio*.

Cimatura - *ital.* also cimature; shearings of wool or cloth.

Cinabro - *ital.* cinnabar (mercuric sulfide HgS) is the chief ore of mercury. The dull brown red ore was mined in Tuscany at Mt. Amiata. It purifies to a brilliant crimson red which was used as a pigment called vermillion, once also known as minium (see *minio*) it is cumulatively poisonous and can ultimately result in organ failure, insanity, and death. In the Roman Empire criminals were sometimes sentenced to work in cinnabar mines in Iberia (now Spain); the average life expectancy of the miners was 3 years. See *mercurio*.

Coagulant - *engl.* see *congulo*.

Coarse salt - *engl.* see *sal grosso*.

Cochineal - *engl.* see *chermisi*.

Colla di pescie - *ital.* literally 'fish glue' or isinglass, or ichtyokolla, a kind of gelatin obtained from various fish, especially sturgeon. In its purest form it is a semitransparent whitish material containing about 80% collagen, obtained by rubbing the fishes presoaked and opened air bladders. A version with lower collagen content is made by boiling fish or eel skin or bones. Egyptians used it from at least 1500 BCE.

Common glass - *engl.* see *vetro commune*.

Compositioner - *engl.* see *conciatore*.

Conciatore - *ital.* glass compositioner. One who makes glass from its constituent materials, adds colorants, and supervises the maintenance and use of the glass melt.

Congulo - *ital.* the alchemical term for the product of coagulation. A solid which has undergone transmutation from a liquid, or in aristotelian terms the element water changing into the element earth. In the light of current chemistry this can best be described by 'precipitation' or 'evaporation'.

Contamination - *engl.* see *immonditia, sporchezza, terrestreità*

Conteria - *ital.* the craft of beadmaking. Although there are many references to beadmaking through out the book, this term was apparently unknown to Neri's first translator, Christopher Merrett, who erroneously translated it as 'counting house' or ignored it entirely. Therefore the term is absent or garbled in subsequent translations, which were all based on Merrett. Etymologically, 'counting house' was a reasonable (albeit nonsensical) guess as the term could well have come from beads used by Catholics to 'count' the rosary as an expression of devotion. See *spiei*.

Conterie - *ital.* glass beads see *conteria*.

Copper - *engl,* see *rame*.

Copper sulfate - *engl,* see *vitriolo di Venere*.

Coreggiolo - *ital.* also correggiuolo,

crogeolo, crociolo, a melting pot or small hand sized crucible sometimes used by goldsmiths, as opposed to the much larger padella and padellotto, which were pots used for the glass melt.

Corpo - *ital.* any solid colored glass formed when an additional ingredient is used in the preparation of an otherwise transparent color to render it opaque. For example in Neri's recipe for rosso in corpo (chapter 58) tin oxide (SnO_2) is used as the opacifier for a blood-red opaque glass formed with Iron and copper oxides.

Correggiuolo - *ital.* see *coreggiolo*.

Credo - *ital.* the apostles' creed, a memorized statement of belief repeated by Catholics as an affirmation of faith in God and the Church. Neri uses it as a timing device in chapter 65 to space out doses of colorant being added to the glass melt. There are shorter versions, but assuming he used the creed of Pius IV, adopted at the council of Trent in 1564; recitation takes a little under 3 minutes (without rushing). See also *Miserere*

Crimson (paint) - *engl,* see *chermisi*.

Cristallino - *ital.* a medium quality glass midway between common glass and crystal.

Cristallo - *ital.* the best quality glass, exceptionally clear and workable.

Cristallo di montagna - *ital.* mountain or rock crystal; a naturally occurring form of exceptionally clear quartz. Neri also uses this term to refer to especially fine crystal glass (artificial rock crystal).

Croco di ferro - *ital.* synonym, see *croco di Marte*.

Croco di Marte - *ital.* literally saffron of Mars, finely ground rust made from oxidized iron filings or through a precipitation of iron (II) sulfate (iron vitriol $FeSO_4$) and alum ($KAl(SO_4)_2 \cdot 12H_2O$), resulting in Fe_2O_3. Also called crocus, crocus of iron, and colcothar. Alternately it may be obtained by heating iron (II) sulfate. When mixed with sal ammoniac (NH_4Cl) and heated it sublimes as iron (III) chloride, which is then converted back into the oxide by ammonia and moisture.

Crocus Martis - *ital.* see *croco di Marte*.

Crocus of iron - *engl,* see *croco di Marte*.

Crocus of Mars - *engl,* see *croco di Marte*.

Crucible - *engl,* see *coreggiolo, padellotto, padella*.

Crude antimony - *engl,* see *antimonio crudo*

Crystal - *engl,* see *cristallo*.

Crystallino - *ital.* see *cristallino*.

Cucchiaio - *ital.* a spoon, or ladle.

Cucchiarata - *ital.* var of cucchiaiata, a spoonful.

Cupric sulfate - *engl,* synonym for copper sulfate, see *vitriolo di Venere*.

Day lily - *engl,* see *fior cappucci*.

Decant - *engl,* see *decanti*.

Decanti - *ital.* decant, to gradually pour from one container into another, typically in order to separate a liquid from its sediment. From Latin *de-* 'away from' + *canthus* 'edge, rim'.

Denaro - *ital.* also danaro. A *pennyweight* is the closest equivalent in the anglo system. A unit of weight currently 1/20 of a troy ounce, but in Neri's time, in Florence 1/24 of a Roman ounce or 24 grains (1.179 modern grams). Usually abbreviated *dwt*, but *pwt* is used in the translation text to distinguish it from the modern unit.

Desalato - *ital*, de-salt, the process by which impurities, and un-dissolved fluxes, especially glass gall, are skimmed off a crucible of molten glass in the furnace.

De-salt - *engl*, see *desalato*.

Destillatoria - *ital*, distillation, the alchemical art of extracting essential oils etc. from plants and herbs. Used in various medicines and remedies.

Diacinto *ital*,- a gemstone referred to by Neri in the introduction to book 5, but not in any of the actual recipes. C. Merrett, his first translator in 1663 translated diacinto as iacinth. (now jacinth or hyacinth). In the Bible jacinth has been supposed to designate the stone called ligure (Hebrew: leshem) mentioned in Exodus 28:19, 39:10-13 as the first stone of the third row in the high priests breast-plate. Also jacinth is mentioned directly in Revelations 21:18-20.
    Properly, the jacinth (or hyacinth) is a flower of a reddish-blue or deep purple and hence a precious stone of that color. From the greek *huakinthos*, with ref. to the mythological story of Hyacinthus, a youth loved, but accidentally killed by Apollo, who caused a flower to grow from the boy's blood.
    Jacinth is currently used to describe the yellow/orange/brown variety of zircon, which when heated can produce a blue stone. To confuse matters further, jacinth/hyacinth has variously been used to also describe sapphire, grossular garnet (essonite) and topaz.

Diaspro - *ital*, jasper a variety of quartz usually colored red or brown by iron oxide, or green by chromium oxide, it is sometimes striated. It polishes nicely and its tight figuring made it highly valued as a gem stone in ancient times.

Dirt - *engl*, see *sporchezza*, *terra*.

Dregs - *engl*, see *greppola*.

Drinking glass - *engl*, see *bicchiere*.

Dry distillation - *engl*, an alchemical technique in which the essence of a material is extracted with dry ingredients only. In the light of current understanding, these were chemical reactions between the ingredients, usually precipitated by ambient heat and or moisture in the air. Neri uses the technique in chapter 31.

Dyestuff - *engl*, see *mori*.

Elixir - *engl*, see *maestra*.

Emerald - *engl*, see **smeraldo**.

Emetites - *ital*, the natural mineral hematite (Fe_2O_3) Chemically this natural mineral is identical to rust. Neri mentions in chapter 125 that the stones were used to shine swords through burnishing. The name derives from the fact that while the polished stone looks metallic, its powder and rubbings on a harder material are bright red, like blood. Greek haimatites (lithos) = 'blood-like (stone)', from haima, haimat- 'blood'. see *spadari*.

Era - *ital*, annealing area or chamber of a glass furnace, a spot where finished work was allowed to cool slowly to avoid stress cracking. Depending on the design of the furnace, this was sometimes in the same chamber as the main glass crucibles, sometimes in another chamber, and sometimes in a separate structure all together. Also called the lehr, lear, or leer.

Esau - *engl, ital*, a biblical character, pronounced 'sw+hairy' was the oldest son of Isaac and Rebecca, and twin brother of Jacob.

Evaporate - *engl*, see *isuapori*.

Eye - *engl*, see *occhio*.

Faua - *ital*, fava, a broad bean rich in potassium, the fava bean is native to the

Mediterranean basin. It is a favorite around the world. The pods are 6 to 12 inches long and the tip of one end is pointed. Inside, there are 5 to 10 flat, round ended seeds which vary in color from green to red to brown to purple. A member of the *vicia* genus, also known as habas, broad bean, faba bean, horse bean, English bean, Windsor bean, tick bean, cold bean, and silkworm bean. Chapter 6 shows a recipe for fava bean crystal.

Fava bean - *engl.* see *faua*.

Felt - *engl.* see *feltro*.

Feltro - *ital.* felt. Used by Neri to filter various liquids. Felt was produced from wool or other natural fibers, which were chemically treated (raising scales on individual fibers) then pressed tightly so the fibers mat and lock together permanently into a sheet of fabric. As a verb the term 'feltro' means to filter (to pass through felt).

Fenugreek - *engl.* see *fieno Greco*.

Ferraccia - *ital.* a metallic tray, in which were placed the lozenges obtained by cutting short sections of glass cane. They were then heated and rounded off in the furnace in order to obtain small 'margarite' beads. Ferracce were also the iron trays that moved along the annealing chamber in order to gradually cool the finished glass work.

Ferret[t]o di Spagna - *ital.* Spanish ferretto, a rich red-brown pigment obtained by calcining copper and sulfur together forming copper oxide Cu_2O which also occurs naturally as cuprite. See chapter 14,15. The term is something of an oxymoron and should not be confused iron oxide (crocus).

Ferretto - *ital.* iron filings. Not to be confused with Spanish ferretto (ferretto di Spagna).

Ferro - *ital.* iron, a metallic element number 26 on the periodic table with a shiny grey luster. The Egyptians were using iron axe heads and chisels before 3000 BCE. In Neri's day blacksmiths were held in great regard for their ornamental wrought iron work, both Florence and Pisa were centers of activity. Neri also uses the term ferro to describe the iron rods used as punties to gather glass from the melt. The island of Elba in the Tyrrhenian Sea was famous for iron mining since Roman times.

Fiandra - *ital.* Flanders, a region that covered the northern half of Belgium, and included parts of France and southern Holland. Neri spent the years between 1604 and 1611, in the Flemish city of Antwerp, sponsored by his friend Emanuel Ximenes, and working in the glass shop of Filippo Ghiridolfi. During this period the country was in the process of regaining independence from Spain. The name Flanders probably is of Celtic origin, derived from the term for 'swampy region'.

Fiaschi - *ital.* flasks, plural of fiasco. This term is most closely associated with the squat, round bottomed, straw covered bottles containing inexpensive wine from the Chianti region. The straw covering helps the bottle sit upright and helps protect the thin glass from breakage. Because this was such a common occurrence, the word has come to be synonymous with sudden disaster, as in the Italian phrase *far fiasco* = 'make a bottle', figuratively means 'fail in a performance'. Also a unit of liquid measure, see *fiasco*.

Fiasco - *ital.* a unit of liquid measure equal to 2.28 liters in Tuscany. See *fiaschi*.

Fieno Greco - *ital.* also fienogreco; the herb fenugreek (*trigonella foenum-graecum*) a 2 to 3 foot tall annual plant with light green leaves and small white flowers, and pods containing 10-20 small flat yellow-brown pungent seeds. Its long history of uses range from Egyptian embalming to flavoring Indian curry, to increasing milk production in breast feeding women, to a coffee substitute, to imitation maple syrup. The bitter seeds are roasted to mellow their flavor and crushed. The leaves are also sometimes used.

Filth - *engl.* see *sporchezza*.

Fior cappucci - *ital.* cabbage flower, a variety of day lily that bears large yellow flowers, each lasting only one day. Genus *hemerocallis*.

Fior de ghiaggivolo - *ital.* see *ghiaggivolo*.

Fioraliso - *ital.* blue iris; fiore+aliso = 'blue lily-flower' (a variety of iris).

Fiori di zolfo - *ital.* flowers of sulfur. A purified form of sulfur produced by sublimation; mineral sulfur is heated, sublimates, and re-condenses on the inside of a glass enclosure. The result is a finely divided, or powdered form of sulfur. The term 'flowers' is an alchemical term for sublimation products in general, it may have its origin in the crystalline floret that are sometimes produced, or it may derive from alchemical symbolism.

Flanders - *engl.* see *Fiandra*.

Flask - *engl.* see *boccia di vetro, fiaschi*.

Flax seed oil - *engl.* see *olio di lino*.

Flint - *engl.* see *pietre focaie*.

Florin - *ital, engl.* from Italian fiorino, diminutive of *fiore* = flower, originally a Florentine coin bearing a fleur-de-lis minted between 1252 and 1553. It was about the size of a US quarter (aprox 1 inch in diameter), and contained 3.54 grams of gold. Adopted in various forms throughout Europe, notably in England, France, Austria, Poland, Germany, Flanders, and Dalmatia.

Flowers of sulfur - *engl.* see *fior di zolfo*.

Flux - *engl.* materials added to silica to lower the melting temperature of glass. These are usually colorless alkali salts, which form oxides and carbonates in the melt. Neri often uses fluxes derived from plant ash high in potassium (K) also containing smaller amounts of Calcium (Ca).

Fornacie di figoli - *ital.* literally a furnace of figures; a potter's kiln in which clay figures and vessels are fired.

Fornacino - *ital.* a small furnace or kiln.

Fornello a vento - *ital.* literally small wind furnace, a furnace or stove fueled by an unforced draft.

Frankincense - *engl.* see *incenso*.

Frit - *engl.* see *fritta*.

Frit kiln - *engl.* see *calcara*.

Frit rake - *engl.* see *riauolo*.

Fritta - *ital.* frit, a mixture of silica and fluxes which is fused in a furnace to make glass. See Neri's description of calcination in chapter 8.

Furnace - *engl.* see *calcara, fornacie, fornacino, fornello*.

Furnace compositioner - *engl.* see *conciatore*.

Garnet - *engl.* see *granato*.

Garofono - *ital.* also garafolo; gillyflower, gilliflower or July flower, any of a number of fragrant flowers, such as the wallflower or white stock. Principally the term is used to refer to the clove (dianthus caryophyllus), of which the carnation is cultivated variety. In Italy it commonly refers to the pink variety. It was used as an alternative to the more expensive oriental clove to spice wines and liqueurs.

In 1662, Neri's first translator Christopher Merrett concludes that fior de ghiaggivolo are gillyflowers; this appears to be in error. Clearly what Neri is referring to (in chapter 110) is the giaggiolo, or the iris, cloves are included here only for completeness.

Gazzera marina - *ital.* the Iberian bird gazza azzurra or azure winged magpie; cyanopica cyanus. It has an intense deep blue plumage on its wings and tail. Not to be confused with the gazza marina, or razorbill, a relative of the penguin also frequents the Italian coast, but sports only black and white plumage.

Gelatin - see *colla di pescie*.

Gesso - *ital., engl.* also gezzo, white plaster, or a mixture of plaster and glue. From the Greek word *gupsos*= gypsum: a soft white or grey mineral consisting of hydrated calcium sulfate ($CaSO_4 \cdot 2H_2O$), used to make plaster of Paris, so called because Paris was a historical an important source of gypsum.

Gezzo - *ital.* see *gesso*.

Gillyflower - *engl.* also gilliflower, see *garofono, ghiaggivolo*.

Ghiaggivolo - *ital.* (fior de ghiaggivolo); the iris, a plant with showy flowers, typically purple or yellow, and sword-shaped leaves. [Genus *iris*: many species.] In 1662, Neri's first translator Christopher Merrett concludes that fior de ghiaggivolo are gillyflowers; this appears to be in error. Clearly what Neri is referring to (in chapter 110) is the giaggiolo, or the iris.

Ginestra - *ital.* broom flower, a shrub typically having many yellow flowers, long, thin green stems, and small or few leaves. *Cytisus scoparius* and many other species, especially in the genera *Cytisus* and *Genista*. The stems were commonly bundled together to fashion various brushes for sweeping.

Girasole - *ital.* girasol, a kind of opal ($SiO_2 \cdot n(H_2O)$) that is essentially colorless and semi-transparent yet retains the shimmering quality of other opals. It has been known since Roman times. Nearly colorless varieties of the red-orange Mexican fire opal are also called girasol, and likewise with moonstone, sunstone, and some colorless sapphires. It homonymously refers to glass with similar properties, usually through the addition of tin oxide (SnO_2) and or bone powder (Hydroxylapatite $Ca_{10}(PO_4)_6(OH)_2$). Some references site it as glass imitating a milky precious stone, which with transmitted light turns a reddish fiery tint. However, Neri uses the term in the context of tinting natural rock crystal (chapter 74) in the color of various red gems. Derived from *gira* = 'to turn' + *sol* = 'the sun'.

Giunchi - *ital.* rush or bull rush, a marsh or waterside plant with slender stem-like pith-filled leaves, some types are used for matting, baskets, etc. Genus *Juncus*.

Glass - *engl.* see *vetro*, 'drinking glass' see *bicchiere*.

Glass compositioner - *engl.* see *conciatore*.

Glass gall - *engl.* see *sale alchali*.

Glaze - *engl.* see *invetriato*.

Golden day lily - *engl.* see *fior cappucci*.

Goslin weed - *engl.* see *robbia*.

Goldsmith's proof - *engl.* see *prova alli orefici, gustare*.

Grain - *engl.* see *grano*.

Granato - *ital.* garnet, a complex group of minerals often cut as gemstones, and running the gamut of color variations (except blue). Chemically they are tri-silicates of the oxides of aluminum, magnesium, calcium, iron, and chromium. Some garnets contain titanium and sodium as well. Hardnesses vary considerably. It was used widely by the Egyptians before 3100 BCE. Neri was probably familiar with the most common form, the deep red pyrope ($Mg_3Al_2(SiO_4)_3$) mined and cut in the Bohemia region (now in the Czech Republic) since about 1500. Another possibility is the orange-red colored alamandine ($Fe_3Al_2(SiO_4)_3$) an ancient source of which was Alabanda in what is now Turkey. In the early 17th century garnet was believed by some to cure depression. The terms granato and ingranato derive from the word for pomegranate, a fruit filled with deep red pulpy seeds resembling the uncut stones.

Gray - *engl.* see *berrettino*.

Grano - *ital.* a grain, a unit of weight equal to 1/24 denaro (denaro = pennyweight or pwt in the text). A Florentine renaissance grain weighed about 0.049 grams. Currently the smallest unit of weight in the troy and avoirdupois systems, equal to 1/5760 of a troy pound and 1/7000 of an avoirdupois pound (approximately 0.0648 grams). From Latin granum; the weight was originally equivalent to that of a grain of wheat.

Green copper - *engl.* see *verde rame*.

Green vitriol - *engl.* see *vitriolo*.

Greppola - *ital.* dregs or dross, in general the solid or semi-solid sediment at the bottom of a liquid. Neri uses the term frequently to indicate the sediment from winemaking. see *tartaro*.

Grisopatio - *ital.* Merrett translates this term to chrysolite, a yellowish green form of the mineral olivine ((Mg-Fe)$_2$-SiO_4). However, other synonyms include chrysoprase, chrysophane, and chrysopal, all of which, happily, can be described as yellow to yellow-green in color.

Gruma - *ital.* also grumma, greppola the raw sediment in the bottom of a bottle or barrel of wine, see *tartaro*.

Gustare - *ital.* literally 'to taste', the process of assaying, or testing a small quantity of material to ascertain its quality or suitability for a given purpose.

Gypsum - *engl.* see *gesso*.

Head - *engl.* see *cappello di vetro*.

Hematite - *engl.* see *emetites*.

Herbal distillation - *engl.* See *destillatoria*.

Homeopathy - *engl.* see *spagirica*.

Hyacinth - *engl.* (or jacinth) see *diacinto*.

Immonditia - *ital.* dirtiness, filth, uncleanliness, contamination. see *sporchezza, terrestreità*.

Impurity - *engl.* see *immonditia, sporchezza, terrestreità*.

Incarnate - *ital.* see *rose incarnate*.

Ingranato - *ital.* see *granato*.

Incense - *engl.* see *incenso*.

Incenso - *ital.* Frankincense, an aromatic gum resin obtained from an African tree (*Boswellia sacra*) and burnt as incense. It is a stimulant although now rarely used as such, Pliny sites it as an antidote for hemlock. Its main use in Neri's time as today is to burn for its aroma.

Invetriato - *ital.* glazed; for many of his enamel recipes (see book 6) Neri uses earthenware containers that have been glazed with a higher melting temperature white glass for various operations.

Iris - *engl.* see *ghiaggivolo*.

Iron crocus - *engl.* see *croco di Marte*.

Iron scales - *engl.* see *scaglia di ferro*.

Isinglass - *engl.* see *cola di pescie*.

Isuapori - *ital.* derived from *suaporare*: to evaporate.

Jacinth - *ital.* (or hyacinth) see *diacinto*.

Jasper - *engl.* see *diaspro*.

Kali - *engl., ital.* salsola kali, also known as saltwort, tumbleweed, and Russian thistle. A shrub bush high in alkali content, common throughout Europe, Asia, and the United States. Uses in the ancient world included heating, soap and glass production. The ash contains 20% K, 18% Ca, 3% Mg, 1.5% Al, 1.5% Fe, 6% phosphate, 6% sulfate, 40% carbonate, and 2% chloride (List and Horhammer, 1969-1979). It is the basis for the modern term alkali, Arabic for from the kali plant. This is probably the source of Neri's rocchetta and polverino.

Kermes - *engl.* see *chermisi*.

Kettle - *engl.* see *caldaro*.

Kettle-smith - *engl.* see *calderaio*.

Kiln - *engl.* see *calcara, fornacie, fornacino, fornello*.

Lacc[h]a - *ital.* lake, or lacquer; an insoluble pigment made by combining a soluble organic dye and an insoluble mordant. Also a purplish-red pigment of this kind, originally one made with lac. See *lacca rossa*. Not to be confused with laghi, lago; a large body of water.

Lacca rossa - *ital.* also laccha, red varnish, lacquer or shellac is a natural resin secreted by the lac insect onto the lac tree of India and Thailand. However it seems likely that Neri, being an astute chemist, would have understood that a purely organic material would reduce to simple carbon ash in the glass melt. It is possible that in his second chalcedony recipe (chapter 43), he was using a different product, or a lac based suspension of red lead (Pb_3O_4), cinnabar (HgS) or colcothar (Fe_2O_3). Lac was imported to Florence from the 13th century, and by Neri's time had become a generic term for a wide variety of red or purplish-red pigments. See *laccha*.

Laccha rossa - *ital.* see *lacca rossa*.

Lacquer - *engl.* see *lacca*.

Ladle - *engl.* see *mestolino, romaiolino*.

Lake - *engl.* see *lacca*.

Lapis emetites - *ital.* see *emetites*.

Lapis lazuli - *ital., engl.* an azure blue gemstone primarily composed of sodium silicate sulfide. Largely mined in what is now Afghanistan, this rock was ground into a light blue-grey powder and subjected to a very laborious process to produce the prohibitively expensive ultramarine pigment. The process essentially resulted in separating out the deep blue lazurite from other constituents. Neri's method is similar to others and is the subject of chapter 115.

Lattimo - *ital.* opaque milk white colored glass. Neri uses tin oxide colorant in his recipes.

Lead - see *piombo*.

Lead monoxide - *engl.* one of several oxide states of lead, PbO occurs naturally in the form of minerals litharge and massicot, both yellow in color. These are so-called dimorphs; minerals with the same chemical formula but different structures, like graphite and diamond. See also *minio*.

Lead salt - *engl.* see *sale di piombo*.

Leek green - *engl.* see *verdeporro*.

Leuante - *ital.* the Eastern Mediterranean region around modern Syria and Lebanon, extending to Greece, Turkey, Israel, and Egypt. 'Levant' loosely translates from French to 'land of the rising sun'.

Levant - *engl.* see *leuante*.

Libra - *ital.* also libbra; a pound, an old Roman unit of weight used in renaissance Florence. In current units it is (339 grams), almost equivalent to ¾ of a modern avoirdupois pound. A Libra then consisted of 12 onces - almost the same ounces we use today. Not to be confused with the Troy pound, which is 373.2 grams.

Lime - *engl.* see *sale della calcine*.

Lime cake - *engl.* see *calcinacci*.

Lime salt - *engl.* see *sale della calcine*.

Limekiln - *engl.* see *calcara*.

Linen - *engl.* see *lino*.

Linen oil - *engl.* see *olio di lino*.

Linguelle di feltro - *ital.* Literally 'tongues of felt' These are filters made of felt, sewn into long conical bags. A bag is suspended over a container, pointing down, and the liquid to be filtered is poured in at the top.

Lino - *ital.* linen; a fabric made from the fibers of the flax plant (*Linum usitatissimum*), highly valued since antiquity as it is generally finer and softer than cotton and other natural fibers. The most prized linens are entirely hand made.

Linseed oil - *engl.* see *olio di lino*.

Litharge - *engl.* see *ritargirio*.

Lixiviation - *engl.* The process of separating soluble from insoluble components by heating a material in liquid, straining, or filtering, then evaporating off the liquid. Neri uses to isolate and purify vegetable salts used as glass flux.

Load - *engl.* see *caricare, scaricare*.

Luna - *ital.* literally moon, an alchemical term for silver, see *argento*.

Lutare - *ital.* lute: To coat with liquid clay or cement to seal a joint, coat a crucible, or protect a vessel from the direct heat of the furnace. From the latin. *Lutum* = 'potter's clay' or mud.

Lute - *engl.* see *lutare*.

Lye - *engl.* see *ranno*.

Macinello - *ital.* a muller; a stone used for grinding various materials such as paint pigments.

Maestra - *ital.* In alchemists' parlance a 'magistery'. A material, potion, or elixir that serves as an agent of change or transformation. The fabled philosopher's stone, purported to cure disease, extend life, and transmute metals was also known as the 'grand magistery'.

Magistery - *engl.* see *maestra*.

Malgama - *ital.* amalgam a compound of mercury (Hg) and another metal, usually silver. From the Greek malagma = emollient. First use of the term has been attributed to Saint Thomas Aquinas.

Mallow - *engl.* see *malva*.

Malva - *ital.* the mallow, a herbaceous plant with hairy stems, pink or purple flowers, and disc-shaped fruit. [Genus *Malva*, family *Malvaceae*: many species.]

Manganese - *ital., engl.* an elemental metal (Mn) atomic number 25 on the periodic chart. Used as a glass decolorant, manganese dioxide (MnO_2) is also known as glassmakers soap. This (the oxide) is the form that was mined in Neri's day; as the mineral pyrolusite, which was crushed to form a black powder. Manganese from the Piedmont region of Italy was especially valued by glassmakers because it contained almost no iron oxide, which was common with product mined elsewhere.

Manganese del piemonte - *ital.* Manganese dioxide (MnO_2) from the

Piedmont region of Italy, especially prized for its low iron content (see manganese).

Mastic - *engl.* see *mastice*.

Mastice - *ital.* mastic, an aromatic gum or resin exuded from the bark of a two Mediterranean trees, used in making varnish and chewing gum and as a flavoring. The small, bushy evergreens, *Pistacia lentiscus*, and *P. cabulica*, both closely related to the pistachio, are the sources of Chios mastic and Bombay mastic respectively. It was highly valued in ancient times for paints, lacquers, adhesives, incense, dental cement, and as chewing gum, from which it derives its name.

Mattone - *ital.* a brick or tile.

Mercurio - *ital.* the elemental metal Mercury (Hg) is number 80 on the periodic chart. It is the only element that is liquid at room temperature. While it is sometimes handled without regard to safety, the pure metal, all its compounds, and especially the vapor is toxic. It accumulates in the brain, liver, kidneys, bones and fatty tissues of the body. Chronic exposure can cause severe neurological disorders, speech problems, autoimmune disease, kidney dysfunction, dementia, and ultimately death. In the Roman Empire, criminals sentenced to work at the cinnabar mines on the Iberian Peninsula (now Spain) had a life expectancy of only three years. Neri uses it liberally in his Chalcedony glass recipes.

Mercury - *engl.* see *mercurio*.

Mestolino - diminutive of mestolo, a small ladle, skimmer, or trowel.

Millet - *engl.* see *saggina*.

Minimum - *engl.* see *minio*.

Minium - *engl.* see *minio*.

Minio - *ital.* minium, or minimum also known as red lead. Lead tetroxide (Pb_3O_4), a bright red or orange-red powder, insoluble in water. As a pigment it has great covering ability and brilliance. Added to the glass melt it becomes colorless. Lead and its compounds are toxic in the same way as mercury, it accumulates in the body eventually resulting in neurological and other irreversible tissue damage. Not to be confused with another brilliant red pigment called vermillion (cinnabar), which is mercury sulfide, HgS. In earlier times, it too was referred to as 'minimum'. The term derives from 'minium river' in north west Spain being the main source of cinnabar for the Roman Empire.

Miserere - *ital.,engl.* the 50th (51st) psalm beginning "Miserere mei, Deus..." (Have mercy upon me, oh God...). Neri uses it as a timing device in chapter 117. Assuming he was using the Vulgate version adopted by Clement VIII in 1592, recital takes about 2 minutes 15 seconds without rushing. See also *credo*.

Mordant - *engl.* see *allume di rocho*.

Mori - *ital.* dyestuff, a substance using or yielding a dye. 'Morire': to dye.

Mortar - *engl.* see *pile di pietra*.

Muller - *engl.* see *macinello*.

Niter - *engl.* see *salnitro*.

Nitric acid - *engl.* see *acqua forte*.

Nitro - *ital.* see *salnitro*.

Occhio - *ital.* the *eye* of the furnace or annealing chamber. This was a hole connecting the combustion chamber where wood was burned to the glassworking areas of the furnace, essentially a heat vent, it was the hottest part of the glass chamber.

Olio di lino - *ital.* literally 'linen oil', this

is flax seed oil, also known as linseed oil. An oil high in omega-3 fatty acids derived by crushing the seeds of the *Linum usitatissimum* plant. The plant's fibrous stems have been valued since antiquity to produce linen fabric, a very soft material, finer than cotton.

Oltramarino - *ital.* an ancient brilliant deep blue pigment made by grinding lapis lazuli into a powder, then through a tedious process separating out the lazurite $(Na,Ca)Al_6Si_6O_{24}(S,SO_4)$ -not to be confused with lazulite. Extremely expensive in the late 16th and early 17th century. From *azzurro oltramarino*, literally 'azure from overseas' (because the lapis lazuli was imported). Neri's separation method is the subject of chapter 115. Ounce for ounce it was more costly than gold.

Once - *ital.* ounce, a unit of weight almost equivalent to a modern avoirdupois ounce. In current units it was 28.25 grams. In renaissance Florence, the Roman system of weights was used in which a libra (pound) was divided into 12 onci, not 16. Not to be confused with Troy weight. See libra.

Opal - *engl.* see *opale*.

Opale - *ital.* opal, a semi-transparent siliceous mineral $SiO_2\text{-}n(H_2O)$ consisting of microscopic spheres of silica, trapping water between them, causing various iridescent rainbow effects and flashes of color to be seen when held in the light.

Opaque - *engl.* see *corpo*.

Opposition - *engl.* an astronomical condition where the earth (in this case) is directly between the moon and the sun; in other words, when the moon is high at midnight. In chapter 5 Neri uses it to describe the best time to harvest ferns for use in crystal glass.

Orange - *engl.* see *rancio*.

Orbio - *ital.* see *occhio*.

Orinale di vetro - *ital.* literally, glass urinal or chamber pot. This was a common and inexpensive glass container in 17th century Europe. Their frequent deployment in Neri's recipes is a prime example of his frugal nature, even while working for Princes, with a fortune in materials.

Orpello - *ital.* gilding or gold paint, however Neri uses the term to describe thin foil strips or tinsel made from copper sheet treated with calamine to form a gold colored brass composition. In chapter 20 he advises saving money by using old wreaths and garlands made from it for a nice blue glass. From the Latin *aurea pellis* = 'skin of gold'. Compare with *orpimento*.

Orpiment - *engl.* see *orpimento*, *orpello*.

Orpimento - *ital.* the bright lemon-yellow poisonous mineral arsenic trisulfide (As_2S_3), formerly used as a dye and artist's pigment as far back as 3100 BCE in Egypt. Now widely replaced by 'king's yellow'. Red orpiment, also known as ruby arsenic, realgar, and as sandarac is arsenic disulfide (As_2S_2) and occurs naturally or is formed by roasting arsenopyrite and iron pyrite together. From Latin '*auripigmentum*' from *aurum* 'gold' + *pigmentum* 'pigment'. Orpiment is moderately toxic if ingested.

Ounce - *engl.* see *once*.

Padella - *ital.* the smaller of the large crucible pots used inside the furnace to hold molten glass.

Padellotto - *ital.* a larger version of the padella, a large crucible used to hold up to several hundred pounds of molten glass in the furnace.

Paioletto - *ital.* also paiuóla, variously a bucket, a flat pan, a kettle, or a cauldron. A cooking vessel, usually metal, with a flat bottom. Diminutive of *paiolo*. A *stagnada* was a copper cauldron.

Palettina - *ital.* a small spoon or spatula used by apothecaries. Diminutive of *paletta* = trowel.

Palm - *engl.* see *porta la palma*.

Pan - *engl.* see *catinella, tegami, teglia*.

Pan-fired beads - *engl.* see *ferraccia*.

Parting water - *engl.* see *acqua partire*.

Pasta - *ital.,engl.* a term Neri uses both to describe hot glass in a pliable putty like state, which for instance may need to be worked on a slab of marble before use, and also to describe frit of glass or enamel which has been ground into powder before it is fused in the furnace.

Paste - *engl.* see *pasta*.

Pastello - *ital.* cake or pie. Also *pasticcio*.

Peach blossom - *engl.* see *perseghino*.

Pearl - *engl.* see *perla*.

Pecie nera - *ital.* black tar. a product of the pine tree and its cones, especially from the Sicilian Aleppo Pine (Pinus halepensis) made by cooking the sap. See also *ragia di pino*.

Pennyweight - *engl.* see *denaro*.

Perla - *ital.* pearl, the distinctive iridescent silvery white color of the natural gem produced by mollusks, which secret 'nacre' around an irritant to produce a small stone-like object prized for its beauty since antiquity. Pearls were extremely popular in ancient China, Egypt, Greece, Rome, and through out Europe in the renaissance.

Perseghino - *ital.* the rich pink color of a peach blossom (not the color of the fruits skin or pulp).

Pestoni - *ital.* Pestle, a heavy tool with a rounded end, used for crushing and grinding substances in a mortar. See *pile*.

Piemonte - *ital.* Piedmont, a mountainous region of northern Italy. Neri calls for the use of manganese oxide specifically mined in the Piedmont region, because of its low iron content. Manganese is used to de-color glass, while iron imparts a green tint, (see *manganese*).

Pietre focaie - *ital.* literally 'fire stones' : flint.

Pignattino - *ital.* a small pot or bowl usually made of clay or ceramic.

Pignatto - *ital.* a pot or pitcher, usually made of clay or ceramic.

Pilatro di leuante - *ital.* also iperico, the herb St. John's wort (Hypericum), A small bushy deciduous tree with five petaled yellow flowers, that grows throughout the temperate regions of the world. Some verities are used today as an herbal remedy for depression.

Pile di pietra - *ital.* Stone mortar a cup-shaped receptacle or bowl in which ingredients are crushed or ground. Neri recommends stone mortars for the preparation of glass ingredients, since iron or bronze would cause unwanted contamination and tint the end product. See *pestoni*.

Pimpernel - *engl.* see *pimpinella*.

Pimpinella - *ital.* pimpernel, a low-growing plant with bright five-petaled flowers. *Anagallis arvensis* (scarlet pimpernel) is one species. From Latin *piper* =pepper. In chapter 33, Neri's reference to 'a wonderful pimpernel green' is presumably alluding to the plant's leaves.

Pine pitch - *engl.* see *ragia di pino*.

Pint - *engl.* see *pinta*.

Pinta - *ital.* a pint, a unit of liquid measure usually used for wine or beer. In chapter 132 of his translation, Merrett writes "six pints of water" for Neri's "libre sei di acqua", changing pounds into pints. The size of a pint varied widely throughout Europe in the 16th and 17th centuries. While it may well have been true for Merrett that an English pint of water weighed about an English pound, this unit was not used in Tuscany, nor by Neri anywhere in his book, but elsewhere in Italy it was equivalent to either a half or a full fiasco (2.089 l).

Piombo - *ital.* the metallic element Lead (Pb), atomic number 82 on the periodic chart. It is a soft bluish grey metal obtained from the mineral galena (PbS). It is an ingredient of many of Neri's glasses, requiring special handling and preparation due to the fact that it drastically lowers the viscosity of the glass and has a tendency to break out the bottom of the crucible in the furnace. Lead and all its compounds are toxic, the vapor and powdered oxides are extremely toxic to breathe. Chronic exposure can lead to hair loss, mental and neurological problems, organ failure, and death. Lead was mined at mount Fiesole in Etruria which overlooks the city of Florence.

Plaster of Paris - *engl.* see *gesso*.

Plinio - *ital.* (the Elder b.23 CE, d.79 CE) Gaius Plinius Secundus, a prolific Roman first century author, naturalist and historian, most noted for his 37 volume Natural History, quoted several times by Neri.

Pliny - *engl.* see *Plinio*.

Poluerino - *ital.* plant ash for glassmaking sold in the form of a powder, hence the name derived from *polvere* = powder, probably produced from the ash of the kali plant. See also *rocchetta*, *kali*.

Polverino - *ital.* see *poluerino*.

Pontil - *engl.* see *canna*, *ferro*.

Poppies - *engl.* see *rosolacci*.

Porfido - *ital.* porphyry, a hard igneous rock containing crystals of feldspar in a fine grained, typically reddish groundmass. Used for grinding and sharpening. From the Greek *porphura* = 'purple'.

Porphyry - *engl.* see *porfido*.

Porta la palma - *ital.* literally 'carries the palm'. An expression denoting 'the best of the best' used by Neri in the title of chapter 35 to describe his very finest green. A reference to the custom of Catholic clergymen carrying palm fronds and leading the Palm Sunday procession, in observance of Jesus' entry to Jerusalem, where his followers laid palm leaves at his feet to walk on. However, the custom of victorious armies carrying palms in a procession through the vanquished territory is one that dates back at least to the early Roman Empire, and possibly much earlier.

Pot - *engl.* see *coreggiolo*, *pignatto*.

Potash - *engl.* see *alum di cantina*.

Pottery kiln - *engl.* see *fornacie di figoli*.

Pound - *engl.* see *libra*.

Prince of Orange - *engl.* see *Principe d'Arangie*.

Principe d'Arangie - *ital.* Maurice, Prince of Orange (1567 - 1625), A brilliant military strategist was instrumental in regaining Flanders' independence from Spain. Technically, his title was Prince of Nassau, until the death of his brother Philip William in 1618. A direct ancestor of the current Dutch Queen Beatrix. He surrounded himself with artists, craftsmen and intellectuals, apparently including Neri.

Proof - *engl.* see *gustare*.

Prova - *ital.* see *gustare*.

Prova alli Orefici - *ital.* Goldsmith's proof; the process of testing a small amount of enamel for color, transparency or consistency by coating a implement made of the metal for which the enamel is intended, which could later be recovered by reheating and quenching the tool in cold water to fracture off the enamel.

Psalm (50th) - *engl.* see *miserére*.

Punty - *engl.* see *canna, ferro*.

Quicklime - *engl.* see *sale della calcina*.

Quocere - *ital.* var. of cócere: to boil, simmer, or bake.

Quocoli - *ital.* very pure quartz river pebbles used in glassmaking. Also called cogli, cougoli, cuocoli, & quogoli.

Ragia di Pino - *ital.* pine pitch, or rosin. The dried sap of the pine tree, the honey colored substance, when cooked becomes black and sticky. It was widely used to waterproof boats, and also as a food flavoring, and as an antibacterial agent. Neri uses it in chapter 115 to make ultramarine blue. see also *pecie nera*.

Rake - *ital. engl.* see *riauolo*.

Rame - *ital.* metallic copper, element number 29 on the periodic chart. Used in ancient (Neolithic) Egypt before 5000 BCE, when it was probably worked cold. By 3000 BCE it was being alloyed with Arsenic, and by 2000 BCE with tin to form Bronze. Brass did not appear until the Roman era 100 BCE. Neri uses the oxides of copper to produce red and green glasses.

Ramina - *ital.* copper scale, flakes of copper made as a byproduct of the smithing process.

Ramina di tre cotte - *ital.* thrice cooked copper = cupric oxide (CuO) occurs naturally as tenorite, and is a black or grey color, which Neri prepares by reducing cuprous oxide (Cu_2O) see chapter 25.

Ramina rossa - *ital.* red copper = cuprous oxide is formed by heating copper, which Neri does indirectly through reverberation (see chapter 24). This is Cu_2O which occurs naturally as the mineral cuprite. A second oxide, cupric oxide (CuO) occurs naturally as tenorite, and is a black or grey color (see *ramina di tre cotte*). Synonym with *ferreto di spagna*.

Rancio - *ital.* also arancio the color orange, the fruit, or the tree; *fior'ranci* = orange blossoms.

Ranno - *ital.* a strong alkali solution called lye, in this case made by boiling vegetable ash. See *lixiviation*.

Receiver - *engl.* see *recipiente*.

Recipiente - *ital.* reveiver, a piece of glassware used by alchemists, essentially a flask that is cemented to the beak of a glass head (alembic) into which condensates of a chemical process are collected.

Red copper - *engl.* see *ramina rossa, ferreto di spagna*.

Red lacquer - *engl.* see *lacca rossa*.

Red lake - *engl.* see *lacca rossa*.

Red lead - *engl.* see *minio*.

Red varnish - *engl.* see *lacca rossa*.

Regis - *ital.* see *acqua regis*.

Retort - *engl.* see *torta*.

Reuerberare - *ital.* to reverberate, a common process in alchemy in which chemical reactions were promoted not by direct heat but through the re-radiation of heat by the furnace walls.

Reverberation - *engl.* see *reuerberare*.

Riauolo - *ital.* a frit rake, as Neri describes it in chapter 2 "the riauolo is an instrument of iron, very long, with which one agitates the frit continuously". This instrument is still very well known in the glass making houses. For the Muranese it remained reàulo until the end of the 1800s: Currently it is known as a reauro.

Rinfocolare - *ital.* to re-fire or reheat, used by Neri in his chalcedony (chapters 42-44) and other glass recipes to indicate a process that glassmakers now refer to as striking. After a piece is worked it is cooled and then reheated, or struck, in a furnace, kiln, or over a torch at a temperature somewhat less than the working temperature. In certain glass compositions, this promotes the growth of microscopic crystals, usually made of reduced metal. A wide variety of colors may be produced, through an effect known as light interference. Exact color is dependant on the size, spacing, and shape of the crystals, which in turn is dependant on glass composition, and the striking conditions.

Ritargirio - *ital.* also litargirio. Litharge, lead monoxide (PbO), A yellow or white form of lead oxides. From Greek lithos 'stone' + arguros 'silver'. Not to be confused with red lead (see *minio*).

Robbia - *ital.* Neri's time the second most important source of dye, after indigo. *Rubia tinctorum* also known as madder root, Turkey red, and alizarin. Its alkaline solution was used with different mordants to give various colors; madder red (with aluminum and tin), blue (with calcium) and violet-black (iron). Purpurin purple dye was also produced with it.

Rocchetta - *ital.* plant ash for glassmaking sold in the form of large pieces, hence the name derived from rocca = rock, probably produced from the ash of the kali plant. See also *polverino, kali*.

Roche alum - *engl.* see *allume di rocho*.

Rock crystal - *engl.* see *cristallo di montagna*.

Romaiolino - *ital.* A ladle. Neri uses glass ladles in chapter 133. See also *mestolino*.

Roman pound - *engl.* see *libra*.

Rose incarnate - *engl.* carnation flowers, literally roses incarnate.

Rosichiero - *ital.* transparent red glass or enamel, specifically created with copper oxide. The name derives from 'rosso+chiara' = clear red. Compare with *rubino*, which is made with gold.

Rosolacci - *ital.* common wild red poppies.

Rosso in corpo - *ital.* literally 'red in body' This is a red glass color that Neri presents in chapter 58. Merrett, translates this to 'deep red'. However, from the context of its use 'opaque' or 'solid' red may be more appropriate.

Rosso piombo - *ital.* Red lead, see *minio*.

Rubino - *engl.* ruby or ruby color. Currently glass called rubino is made with a colloidal suspension of microscopic crystals of gold. Neri was apparently aware of the use of gold as a basis for red glass, which he alludes to in chapter 90, and then presents a short recipe for a rubino glass in chapter 129, apparently to be used as an enamel. Kunckel describes a rubino glass made with gold in the annotations to his 1679 German translation of L'Arte Vetraria although he holds back the exact details of the recipe. Also see *rosichiero*.

Ruby - *engl.* see *rubino*.

Rush - *engl.* see *giunchi*.

Saggina - *ital.* a grass, or cereal which bears a large crop of small seeds, used to make flour or alcoholic drinks. Includes *panicum miliaceum* and other species.

Saint Jerome - *engl.* see *San Girolamo*.

Saint John's wort - *engl.* see *pilatro di leuante*.

Sal ammoniac - *engl.* see *sale ammoniaco*.

Salci - *ital.* willow trees.

Sale - *engl.* salt, in this context Neri is referring to the vegetable salts extracted from plant ash, through lixiviation, mostly potassium and calcium compounds. In general it is any chemical compound formed by the reaction of an acid with a base, with the hydrogen of the acid replaced by a metal or other cation.

Sale alchali - *ital.* alkali salt, also known as glass gall, sandever, and sandiver. The (foul smelling) whitish salt skimmed from the surface of melted glass, which is cast up, as a scum, from the fusion process. It is essentially excess salts and other impurities in the frit.

Sale ammoniaco - *ital.* also sale armoniacó. Sal ammoniac = ammonium chloride NH_4Cl. A white crystalline powder added to nitric acid to form aqua regia. In ancient Egypt the priests at the temple of Amun distilled soot. The 'Salt of Amun' thus obtained was called by the Romans *sal ammoniac*.

Sale armoniacó - *ital.* see *sale ammoniaco*.

Sale della calcina - *ital.* lime salt, calcium oxide (CaO), also known as 'quicklime' a caustic alkaline white powder made by heating limestone or seashells ($CaCO_3$) in a process invented by the Romans. Used in the manufacture of cement for the construction trades.

Sale di piombo - *ital.* lead salt, which in chapter 91 Neri extracts from ceruse, in all probability this snow white material is mainly a mixture of lead carbonate in its hydroxide form, $2PbCO_3 * Pb(OH)_2$ and its base carbonate $PbCO_3$. It may well also contain significant amounts of lead acetate $Pb(C_2H_3O_2)_2$ which occurs in many chemical variations, and lead hydroxides $Pb(OH)_2$, $Pb_2O(OH)_2$. The oxide states of lead include PbO, which is white, in addition to others that can be red, yellow, black, or grey in color. Ultimately, all of these compounds decompose at relatively low furnace temperatures, so they are not an indication of the final composition of Neri's glasses beyond their basic lead content.

Sale di Saturno - *ital.* see *sale di piombo*.

Sale grosso - *ital.* coarse salt.

Sale marino - *ital.* sea salt, see *sal nero*.

Salina bianca - *ital.* white salt: sodium chloride, or table salt. In chapter 36 Neri sites Voltera as a location for its production.

Sal nero - *ital.* black salt. Some references site a sea salt, which is mainly composed of sodium chloride ($NaCl$), others describe it as mainly potassium chloride (KCl) mined heavily in the orient, often with a sulfurous smell due to other components in varying amounts which can include compounds or salts of sulfur, iron, zinc, nickel, magnesium, manganese, copper, titanium, calcium and sodium . In chapter 36 Neri uses it as an opacifier for sky blue glass.

Salnitro - *ital.* saltpeter, also called niter. A mineral composed of potassium nitrate KNO_3. A colorless crystal or white powder. It has a sharp saline taste and is soluble in water. It was used in the manufacture of dyes, and Neri uses it in the production of nitric acid (aqua fortis) in chapter 38. Natural formations occurred in arid regions as efflorescence formations on rocks and masonry. It was also produced throughout the middle ages and the renaissance by fermenting (oxidizing) animal dung, urine, and plant matter together with lime in open troughs. Some references confuse niter with natron ($Na_2CO_3 \cdot 10(H_2O)$), and sometimes it is used as a synonym for $NaNO_3$.

Salt - *engl.* see *sale*.

Saltpeter - *engl.* see *salnitro*.

Saltwort - *engl.* see *kali*.

San Girolamo - *ital.* (347 CE - 419 CE) Saint Jerome was a controversial and prolific late fourth century theologian. He wrote a revision of the Latin version of the Book of Job in the year 384, and authored the Vulgate translation of the Old Testament. Reputedly, he removed a thorn from a lion's paw, thereby winning its devotion for years. Also known as Eusebius Hieronymus Sophronius, and Hieronymus.

Sapphire - *engl.* see *zaffiro*.

Saturn - *engl.* see *saturno*.

Saturno - *ital.* Saturn, an alchemical term for lead.

Saxon blue - *engl.* see *zaffera*.

Sbiadato - *ital.* a faded or washed out color, but in Neri's time also sometimes used to denote a light sky blue.

Scaglia di ferro - *ital.* iron scales, which flake off of a piece of red-hot iron when it is pounded with a hammer on an anvil. For Neri this was a low cost source of iron filings.

Scaricare - *ital.* to unload; Neri uses this term to describe the process of lowering the pigmentation of a batch of glass that has had too much colorant added. See *caricare*.

Scodella - *ital.* a bowl. See also *pignattino*.

Sea salt - *engl.* see *sale marino*.

Sediment - *engl.* see *sporchezza, terrestreità*.

Sermenti - *ital.* buds, sprigs, sprouts or young branches of a vine. Also water reeds.

Shearing - *engl.* see *cimatura*.

Sieve - *engl.* see *staccietto, tamigrato*.

Silver - *engl.* see *argento*.

Smalto - *ital.* enamel, in glassmaking this was a pigmented glass powder used to cover gold, silver or glass articles. Sometimes used to paint colorful scenes or for lettering. It was applies in thin layers, sometimes mixed with a liquid binder, dried, and fired in a kiln or over an oil lamp. This process was repeated until the desired effect was achieved. Finally, the piece was polished smooth.

Smalto azzurro - *ital.* a blue paint pigment that Neri uses as an ingredient but does not explain further. Copper carbonate $CuCO_3$ is one possibility. It was used widely as an economical alternative to the prohibitively expensive ultramarine. Copper carbonate is essentially the artificially produced blue component in the mineral azurite $2(CuCO_3) \cdot Cu(OH)_2$. Another possibility is that Neri's *smalto azzurro* is a cobalt oxide based product (see *zaffera*), which is the theory favored by his first translator Merrett (1661, p.320). The term 'smalts' was used in the glass and metals crafts to indicate a thin colored glass coating over metal (enamel), while painters and fine artists used it to indicate finely ground glass (often cobalt blue) used as a pigment. It seems likely that Neri would have known this.

Smeraldino - *ital.* the rich green color of the emerald gem.

Smeraldo - *ital.* emerald, a gemstone known in ancient times, mined in Egypt before 2000 BCE and once a favorite of Cleopatra, these stones are in the beryl family $Be(Al,Cr)_2\text{-}Si_6O_{18}$. In the 16th century, the Spanish conquest resulted in huge quantities of emeralds imported to Europe from Columbia in South America. Emeralds derive their distinctive color from trace amounts of chromium and vanadium.

Soda - *ital., engl.* used by Neri as a generic term for plant ash from a variety of sources. From the Arabic term *suwwad* meaning 'saltwort'(kali). In current usage, it refers to Sodium carbonate Na_2CO_3.

Soda di Spagna - *ital.* plant ash for glassmaking from Spain.

Solid (color) - *engl.* see *corpo*.

Spadari - *ital.* swordmakers, plural of 'spadaio', an artisan skilled in the ancient craft of making swords and knives. In chapter 125, Neri refers to hematite (Fe_2O_3) with which swordmakers use to burnish. The polished stones were used to shine and darken the surface of blades by rubbing. Powdered material was also used to polish blades. These stones were also used in the gold gilding process, since gold does not stick to hematite. see *emetites*.

Spagirica - *ital.* the art of reducing things to their constituent elements (or essences), purifying them, and then reconstituting them. This was accomplished by a variety of means including solution, distillation, and evaporation.

Spagyric - *ital.* see *spagirica*.

Spanish barilla - *engl.* see *soda de Spagna*, and *barilla*.

Spanish ferreto - *engl.* see *ferreto di Spagna*.

Spanish soda - *engl.* see *soda di Spagna*.

Spatula - *engl.* see *palettina*.

Spiei - *ital.* from spiedo, spedo, spedare; to roast on a spit, or to skewer. A term used to indicate the process through which sections of large diameter pierced glass cane were rounded in the furnace 'rotisserie style' on a rotating iron mandrel. This is how rosary beads = *paternostri* (literally 'our fathers') and *conterie* were obtained.

Spirit of Saturn - *engl.* see *Anima di Saturno, Zucchero di Saturno*.

Spit formed beads - *engl.* see *spiei*.

Spoon - *engl.* see *cucchiaio, palettina*.

Spoonful - *engl.* see *cucchiarata*.

Sporchezza - *ital.* dirt, filth, foulness, impurity. Derived from the word porcile: pigsty. See also *immonditia, terrestreità*.

Staccietto - *ital.* a fine sieve, See *staccio*.

Staccio - *ital.* a sieve, a utensil consisting of a screen or mesh held in a frame, used for straining solids from liquids, for separating coarser from finer particles, or for reducing soft solids to a pulp. See *staccietto, tamigrato*.

Stagno - *ital.* the metal tin. It has a silver-white lustrous appearance with a bluish tinge. It is soft and malleable. The principal source of tin is the mineral cassiterite (SnO_2). Small amounts of SnO_2 dramatically increase the fluidity and luster of glasses. It is frequently used as an opacifier in enamels and pottery glazes. Tin's other oxide, SnO, when mixed with magnesium and cobalt oxides produces the painters pigment cerulean blue. Mixed with copper oxide it produces a ruby red glass. Pewter of the Roman Empire consisted of about 70% tin and 30% lead, but in Neri's time pewters were made with over 90% tin.

Steel - *engl.* see *acciaio*.

Stirring rod - engl. see *bastoncino*.

Stort - ital. a retort, a glass container with a long neck bent to one side and downward, used in distilling liquids and in other chemical operations. Liquids are heated in the body of the retort, evaporate, condense at its top, and run down the neck and out the spout.

Strike - engl. see *rinfocolare*

Suaporare - ital. also isuapori: to evaporate.

Sublimation - engl. see *sublimatione*.

Sublimatione - ital. sublimation, the conversion of a substance from a solid directly into a vapor, usually with the application of heat, and then often back again into a solid. This was a basic alchemical process used in the purification of various substances. The alchemical term for the products of sublimation was 'flowers'; flowers of sulfur, flowers of zinc, etc. perhaps due to the crystal florets produced.

Sugar of Saturn - engl. see *zucchero di Saturno*.

Sulfur - engl. see *zolfo*.

Sulfur anhydride - engl. see *vitriolo*.

Sulfur of Saturn - engl. see *zolfo di Saturno*.

Sulfuric acid - engl. see *vitriolo*.

Suolo di zolfo - ital. literally *sulfur from the ground* = brimstone.

Swordmakers - engl. see *spadari*, (singular: spadaio).

Tamigrato - ital. to sift or strain. tamigio is a sieve or strainer. see also *staccietto*.

Tarso - ital. made from quartz pebbles (quocoli) of exceptionally pure silica (SiO_2) found in riverbeds, which were pulverized to form silica powder, which is a basic ingredient for all Neri's glasses. The Muranese preferred stones from the Ticino River in Pavia.

Tartar - engl. see *tartaro*.

Tartaro - ital. tartar, also known as gruma greppola and argol. The reddish incrustation that forms on the inside walls of aged red wine barrels consisting of yeast mixed with potassium bitartrate. Chemically it is a potassium compound formed through a reaction with tartaric acid, a major constituent of grape juice. Pure tartar takes the form $KHC_4H_4O_6$. Neri prefers to obtain his material from casks of red wine, and warns against the product available in powdered form (although he ultimately powders it himself).

Tegame - ital. a pan, plural= tegami. In chapter 131 Neri describes it as a "terracotta vessel of round shape, and flat bottom, that in Tuscany are called tegami". See also *catinella, teglia*.

Tegamino - ital. a small shallow pan (in this case probably terra cotta), diminutive of tegami.

Teglia - ital. a low flat pan. see also *catinella, tegami*.

Térra - ital. also terra; earth, clay, dirt. Terracotta is fired clay or pottery.

Terrestreità - ital. sediment, earth, dirt. That which has the essence of the earth. See also *immonditia, sporchezza*.

Test - engl. see *gustare*.

Thrice cooked copper - engl. see *ramina di tre cotte*.

Tiberio - ital. Tiberius Claudius Nero (42 BCE to 37 CE) followed Augustus as emperor of Rome. Pliny and Petronius recorded the story of an artisan who upon discovering a flexible glass was put to death by Tiberius in order to protect the value of his gold and silver holdings.

Tiberius - *engl.* see *Tiberio*.

Tile - *engl.* see *mattone*.

Tin - *engl.* see *stagno*.

Tinsel - *engl.* see *orpello*.

Topaccio - *ital.* rich golden brown topaz color.

Topatio - *ital.* the gemstone topaz, (Al_2SiO_4 $(Fe,OH)_3$) commonly has a golden brown to yellow color and is sometimes confused with citrine. Natural blue and red stones do exist but are extremely rare. Imperial topaz, a sherry colored variety is the one historically associated with the name since the Egyptian and Roman empires. Imperial topaz stones sport colors ranging from brownish yellow to orange yellow to reddish brown. It seems possible from the context (chapter 74) that Neri was referring to this variety.

Topaz - *engl.* see *topatio*.

Tongue (of felt) - *engl.* see *linguelle di feltro*.

Torta - *ital.* retort, *see stort*.

Trag[i]ettare - *ital.* a term Neri uses for the process of throwing or flinging molten glass into vats of water as a means of removing excess salt, and other impurities (glass gall). He also utilizes this method to remove excess metallic lead from his lead glasses.

Transparent Red - *engl.* see *rosichiero*.

Tray - *engl.* see *catinella*.

Trebiano - *ital.* also trebbiano, a white wine native to Italy, possibly the 'Trebulanum' described by Pliny. The derivation is from 'trebbiare': to thresh. Italy's most widely planted vine, and France's most widely planted white vine. It is also the most common base for Cognac and Armagnac, being high in both acid and alcohol content.

Trementia - *ital.* turpentine, also oil of turpentine. The distillation of an oleoresin secreted by certain pines and other trees. A volatile pungent oil, used in mixing paints and varnishes and in liniment. From the Latin. ter(e)binthina (resina) = '(resin) of the terebinth'.

Troy - A system of weights used for precious metals and gemstones based on a 12 oz pound and a 31.10 grams. See *avoirdupois*.

Turpentine - *engl.* see *trementia*.

Ultramarine - *engl.* see *oltramarino*.

Unload - *engl.* see *scaricare, caricare*.

Urinal - *engl.* see *orinale di vetro*.

Verde rame - *ital.* also vederame literally 'green copper', Merrett translates this into Verdigris = copper acetate. the hydrate, $Cu(CH_3COO)_2 \cdot H_2O$, was known as crystals of Venus. It was used widely as a (poisonous) paint pigment. However, the mineral Malachite $CuCO_3 \cdot Cu(OH)_2$ or green copper ore was also widely used, and is a possibility here. Heating copper acetate forms a glacial acetic acid. $2Cu(CH_3CO_2)_2 = 2Cu + 3CH_3CO_2H + CO_2 + C$

Verdeporro - *ital.* a bright green color of the of the leek, a plant related to the onion, with flat overlapping leaves forming an elongated cylindrical bulb that together with the leaf bases is eaten as a vegetable. *Allium porrum*.

Verdigris - *ital.* see *verde rame*.

Vermillion - *ital.* see *minio*.

Vertificare - *ital.* to convert into glass or a glass-like material through the application of heat.

Verzino - *ital.* the wood of the Brazilian trees *Caesalpinia brasiliensis, C. crista*, and *C. echinata*, used for dyes. It produces purple shades with chrome mordant, and crimson with alum. The wood is prized for fine furniture and violins; it has a rich bright-red color, and takes a fine lustrous polish. Also known as Brazilwood.

Vetriolo - *ital.* see *vitriolo*.

Vetro - *ital.* glass. Drinking glass see *bicchiere*.

Vetro commune - *ital.* common glass, the most basic and lowest quality glass.

Vine twigs - *engl.* see *sermenti*.

Vinegar - *engl.* see *aceto*.

Viper color - *engl.* see *vipera*.

Vipera (colore) - *ital.* the color of the viper snake. The common viper = adder snake (*vipera beris*) is found through out the Italian peninsula, as well as the greater part of Europe, and is by far the most common viper. The female can be a russet red color with orange yellowish highlights usually with a distinctive black zigzag pattern running down its back. The male snake is monochrome.

Vitrification - *engl.* see *vertificare*.

Vitriol - *engl.* see *vitriolo*.

Vitriol of copper - *engl.* see *vitriolo di Venere*.

Vitriol of Venus - *engl.* see *vitriolo di Venere*.

Vitriolo - *ital.* Various metal oxide states of sulfur (historically either 'green vitriol' (FeSO4) of iron, or 'blue vitriol' (CuSO4) of copper.) The two have distinctly different chemical properties, however it is clear in L'Arte Vetraria that Neri consistently uses 'blue vitriol' or copper sulfate, which he describes in chapters 31, 39, & 131). When dissolved in water, heated and then allowed to oxidize in air, both vitriols react to form an impure sulfuric acid (H_2SO_4). Copper sulfate was also used by some (e.g.,Geo. Agricola) to produce nitric acid; however, it does not appear in Neri's formulation. From the Latin vitrum = glass, due to the lustrous minerals (pyrites) from which it is derived. See also *Vitriolo di Venere*.

Vitriolo di rame - *ital.* synonymous with vitriolo di Venere, Venere (Venus) was an alchemist's term for copper, and its vitriol was copper sulfate. see *vitriolo*.

Vitriolo di Venere - *ital.* vitriol of Venus. Cupric sulfate ($CuSO_4$) also called bluestone. Naturally occurring as Chalcanthite, this is a synonym for *vitriol di rame*. See also *vitriolo*.

Wax - *engl.* see *cera*.

White Bronze - *engl.* see *acciaio*.

White Lead - *engl.* see *cerusa*.

White salt - *engl.* see *salina bianca*.

Willow - *engl.* see *salci*.

Wind furnace - *engl.* see *fornello a vento*.

Wine dregs - *engl.* see *gruma, greppola, tartaro*.

Zaffer - *engl.* see *zaffera*.

Zaffera - *ital.* Saxon blue, a mineral mixture produced in Saxony. A deep-blue powder made by fusing cobalt oxide with silica and potassium carbonate, It contains 65 to 71% silica, 16 to 21 potash, 6 to 7 cobalt oxide, and a little alumina.

Zaffiro - *ital.* the gem sapphire, a form of corundum Al_2O_3 occurs in a wide variety of colors, the most popular being blue. Red sapphire is ruby. The coloration is caused by trace metal ions trapped in the crystal matrix. It is likely that for Neri, sapphire was blue, as that color gem has been associated with the name since antiquity.

Zanech - *ital., engl.* Neri sites him as the father of Job according to Saint Jerome. This is not supported in current biblical scholarship. It is possibly a reference to Hanoch, grandson of Abraham & Keturah (Abraham-Midian-Hanoch) as distinct from the Abraham & Sarah bloodline (Abraham-Isaac-Esau).

Zelamina - *ital.* calamine a mineral ore occurring in crystal groups with a vitreous luster. It may be white, yellowish, greenish, or brown. Chemically it is a zinc silicate ($2ZnO \cdot SiO_2 \cdot H2O$) usually only containing about 3% zinc. It was concentrated through roasting and distillation, and then mixed directly with copper to form Brass. This mineral was often associated with the weathering zones around silver mines. (Not to be confused with the soothing pink ointment called calamine, which is made from a mixture of zinc carbonate and ferric oxide.)

Zolfo - *ital.* the element sulfur (S), also known as Brimstone, sulfur is number 16 on the periodic table and is mined in its pure lemon yellow crystalline state in Sicily. Flowers of sulfur (see Fiori di zolfo) was used for a variety of purposes including as a disinfectant from before 2000 BCE. Neri used it in the production of sulfuric acid and related compounds. See *vitriolo*.

Zolfo di Saturno - *ital.* sulfur of Saturn, a finely powdered version of sugar of Saturn. See *zucchero di Saturno*.

Zucchero di Saturno - *ital.* sugar of Saturn, a purified form of what is probably lead acetate, with amounts of carbonate and hydroxide. see sale di piombo. It was noticed as early as the Roman Empire that bitter or turned wines could be sweetened by the addition of ceruse (lead carbonate) and other lead compounds which dissolve in the wine's acetic acid to produce sweet tasting, but deadly lead acetates. See also *Anima di Saturno*.

BIOGRAPHY

Appearing below are the names of people pertinent to *L'Arte Vetraria*, and Neri's life. Included are direct references made in his book, those who, from other sources we know influenced his life, as well as important translators and biographers.

Andrea - son of Antonio Neri, along with Fillippo, Pierantonio, and Francesco, according to a genealogical record at the State Archives of Florence by Alberi Pucci: Genealogici. (Not corroborated by a second source.)

Bartolini, Alamanno - Employer of Antonio Neri, brother in law to Emanuel Ximenes and husband of Beatrice. His palace at S. Trinita Square is still standing in Florence.

Bartolini, Beatrice - Ximenes sister, wife of Alamanno.

Buontalenti, Bernardo - (1536 - 1608) A principal architect for the Medici family under Cosimo, Francesco, and Ferdinando. He also served as military engineer, drawing master, fireworks display designer, porcelain decorator, and was a glassworker in his own right. He was designer of the Casino palace, where Antonio worked at the start of his career.

Cappello, Bianca - (1548 - 1587) Adoptive mother of Don Antonio Medici, long time mistress and second wife of Grand Duke Francesco (after Joanna). Both Bianca and Francesco died together suddenly (from malaria) when don Antonio was eleven.

Cesi, Federico - (1585-1630) founder of the scientific 'Society of the Lynx-eyed', recommended L'Arte Vetraria to friend Galileo Galilei in a letter dated 1614.

Dianora - Antonio Neri's mother, maiden name Francescho. Married 10 August 1570 to Neri Jacopo. Mother to Jacopo ('73), Francesco ('75), Antonio ('76), Jacopo ('77), and Vincenzo ('79).

Esau - a biblical character, pronounced 'sw+hairy' was the oldest son of Isaac and Rebecca, and twin brother of Jacob.

Federigho Sassetti, Ginevera - Antonio Neri's godmother.

Fillippo - son of Antonio Neri, along with Pierantonio, Andrea, and Francesco according to genealogical record State Archives of Florence: Alberi Pucci, Genealogici. (Not corroborated by a second source.)

Francescho, Mr. - Antonio Neri's maternal grandfather, a Florentine lawyer.

Francesco - son of Antonio Neri, along with Fillippo, Pierantonio, and Andrea according to genealogical record State Archives of Florence: Alberi Pucci, Genealogici. (Not corroborated by a second source.)

Galileo, Galilei - (1564-1642) Astronomer, mathematician, physicist, In a letter of 1614 thanked friend F. Cesi for recommending it commenting that it was "very rich in practical knowledge and beautiful artistry."

Ghiridolfi, Filippo - Ran the Antwerp glass shop where in 1609 Neri demonstrated "...the most beautiful chalcedony glass that I have ever made in my life."

Grand Dukes - see Cosimo Medici, Francesco Medici, Ferdinando Medici.

Hollandus, Isaac - see Isach Olando.

Jacopo - Antonio's paternal grandfather.

Jacopo, Neri - Antonio's father, a practicing physician, son of Jacopo, married Dianora 10 August 1570. Father to Jacopo ('73), Francesco ('75), Antonio ('76), Jacopo ('77), and Vincenzo ('79). The name Neri is possibly a shortened version of Raneri.

Kunckel, Johan - (cir.1630 - 1703) Alchemist, son of a master glassmaker. cir 1667-77 worked in the service of Johan Georg II, Elector of Saxony at Dresden. 1679-1688 worked in service to Frederic William, Elector of Brandenburg. Charged with operating chemical and glassworks, in Berlin and Potsdam. In 1686 erected crystal glass works on Pfaueninsel Island near Potsdam, succeeded in manufacturing Rubino glass, 1679 published German translation of *L'Arte Vetraria*. Lost favor with the Prussian court after the death of his patron. In 1689 the island laboratory was burned to the ground. In 1693 conferred title of Barron Kunckel von Löwenstern by King Charles XI of Sweeden.

Landi, Niccoló - Scheduled furnace time at the Medici Casino glasshouse, and a good friend of Antonio Neri.

Medici, Cosimo I - (1519-1574) Father of Francesco and Ferdinando.

Medici, Cosimo II - (1590 - 1621) son of Ferdinando, acceded in 1609. Patron of Galileo.

Medici, Don Antonio - (1576-1621), Patron of Antonio Neri. The son of commoners but passed as the offspring of Bianca Cappello and Grand Duke Francesco de Medici. Was raised like a prince in the Medici household and treated as such even after the truth of his origin was known (but was conferred with the cross of the Knights of Malta, in order to ensure his celibacy). He made his residence in Florence, at the Medici palace called the 'Casino' located between Saint Mark's Square and Via San Gallo. There he was trained by alchemists, spending, as one historian from Tuscany wrote two centuries ago, "immense sums of gold in order to learn and to experience various secrets that were sold to him by imposters at a dear price", but succeeding also "to collect and to verify a great number of secrets pertaining to medicine, and to the perfection of diverse arts" (cf. *Notize de G. Taragoni Tozzetti*; Florence, 1852 p 256).

Medici, Ferdinando - (1549-1609) son of Cosimo I, brother of Francesco, 'uncle' to don Antonio. Left his post as Cardinal in Rome to become third Grand Duke of Tuscany when his brother died in 1587. He was responsible for renovating the glass facility in Pisa, where Neri worked. He was succeeded as Grand Duke by his son Cosimo II.

Medici, Francesco - (1541-1587) Adoptive father of Don Antonio, husband to Bianca Capello. Second Grand Duke of Tuscany in 1564 upon the death of his father Cosimo I. He was brother to Ferdinando, of whom he was not fond. With his first wife Joanna he had seven children, including one boy, Filippo who died at age five. He spent a large part of his private time indulging his passion for alchemy. He had the Casino di San Marco build around 1575.

Merrett, Christopher - (1614-1695) First translator of L'Arte Vetraria (1661), friend of Robert Boyle. Oxford University educated, Physician, an original member of the Royal Society, formed in 1660. Keeper of the library of the Royal College of Physicians, 1565-6. 1581 expelled from same.

Neri, Antonio Lodovicho - (1576-1614?) author of *L'arte Vetraria*, son of Dianora and Neri Jacopo. Brother to Jacopo, Vincenzo, Francesco, and another Jacopo.

Neri, Francesco - (1575 - ?) Antonio's brother along with Vincenzo, and two Jacopos. Son of Neri and Dianora.

Neri, Jacopo - (1573- ?) Antonio's brother along with Vincenzo, Francesco, and yet another Jacopo. Son of Jacopo and Dianora.

Neri, Jacopo - (1577-?) Antonio's brother along with Vincenzo, Francesco, and yet another Jacopo. Son of Jacopo and Dianora.

Neri, Vicenzo - (1579 - ?) Antonio's younger brother, holding the post of Friar in the church. Son of Jacopo and Dianora.

Olando, Isach - Also known as (Johannes) Isaac Hollandus. It is unclear whether *Johannes* is part of the same name or that of a second person (sibling/child). He has been credited with early development of enameling precious metals. In his book '*Opera Mineralia...*' he describes the use of metal oxides to color stones and crystals, and to make imitation gems. He was apparently a German, Swiss, or Dutch alchemist, and although legend and pseudononymous authors have confused the issue of his lifetime, placed it sometime between the 13th and 17th centuries, it seems clear that the author if Neri's reference almost certainly belongs to the 16th century.

Pierantonio - Son of Antonio Neri, along with, Fillippo, Andrea, and Francesco according to genealogical record State Archives of Florence: Alberi Pucci, Genealogici. (Not corroborated by a second source.)

Piero, Mr. - Florentine cannon lawyer, appeared as a witness for Vincenzo Neri's birth.

Plinio - the Elder (23-79) Gaius Plinius Secundus, a prolific first century Roman author, naturalist and historian, most noted for his 37 volume Natural

Principe d'Arangie - (1567 - 1625) Maurice, Prince of Orange, A brilliant military strategist was instrumental in regaining Flanders' independence from Spain. Technically, his title was Prince of Nassau, until the death of his brother Philip William in 1618. A direct ancestor of the current Dutch Queen Beatrix. He surrounded himself with artists, craftsmen and intellectuals, apparently including Neri.

Saint Jerome - see San Girolamo.

San Girolamo - (347-419) Saint Jerome was a controversial and prolific late fourth century theologian. He wrote a revision of the Latin version of the Book of Job in the year 384, and authored the Vulgate translation of the Old Testament. Reputedly, he removed a thorn from a lion's paw, thereby winning its devotion for years. Also known as Eusebius Hieronymus Sophronius, and Hieronymus.

Sisti, Niccoló - Operator of Pisa glass and ceramics facility where Antonio Neri worked in the early 17th century, in the courtyard of what is now 43-44 Lungarno.

Tiberio - (42 BCE-37 CE) Tiberius Claudius Nero followed Augustus as emperor of Rome. Pliny and Petronius recorded the story of an artisan who upon discovering a flexible glass was put to death by Tiberius in order to protect the value of his gold and silver holdings.

Tiberius - see Tiberio.

Ximenes, Emanuel - Brother of Beatrice, friend of Antonio Neri, hosting him on his long stay in Antwerp between 1603- 1611.

Zecchin, Luigi - (1905-1984) Muranese glass scholar, and arguably Neri's first serious biographer.

"Above all is this wonderful invention. A new way practiced by me, with the doctrine taken from Isaac Hollandus, in which paste jewels of so much grace, beauty, and perfection are made, that they seem nearly impossible to describe, and hard to believe."

- Antonio Neri chap. 75

SELECTED BIBLIOGRAPHY

Abbri, Ferdinando, *Antonio Neri, L'Arte Vetraria*. Florence: Giunti, 2001, ISBN 88-09-01267-4 (paper). Italian reprint by Neri's original publisher, with updated spelling and typography.

Ball, Philip, *Bright Earth, Art and the Invention of Color*. New York: Farrar, Straus, and Grioux, 2001, ISBN 0-374-11679-2. The story of paint pigments and their contribution to the development of modern chemistry. A thoughtful and compelling read.

Barovier, Rosa (Mentasti), *L'Arte Vetraria by Antonio Neri*. Milan: Polifilo, 1980, ISBN 88-7050-404-2 (cloth). Introduction in both English and Italian. This, as far as I know, is the only reprint which is a direct verbatim facsimile of an original 1612 printing. It is out of print but still available through some Italian book dealers.

Burckhardt, Jacob, *The Civilization of the Renaissance in Italy*. New York: Barnes and Noble, 1999, ISBN 0-7607-1545-9. Reprint of the 1860 classic. This is ground zero for English language history of Renaissance Italy, a must for anyone seriously interested in the subject.

Catholic University of America, *New Catholic Encyclopedia*, Farmington Hills, MI: Gale Group, 2002. Weighing in at 15 volumes, this is out of range for most individuals, but not university libraries. The original 1913 edition is quite informative and on the web at www.newadvent.org (not affiliated with Gale).

Coyne, G. V., et al. *Gregorian Reform of the Calendar*. Rome: Specola Vaticana, 1983. This is the definitive work on the subject; proceedings of the Vatican conference commemorating 400 years of the Gregorian system. Published in English and Italian versions.

Florio, John, *A Worlde of Wordes*, New York: Gerog Olms Verlag, 1972, ISBN 3487042274, (ed: Bernhard Fabian, et al.) Reprint of the 1598 Italian to English dictionary, out of print and hard to find, but all 480 pages may be downloaded for free from the French National Library website: www.bnf.fr

Freedberg, David, *The Eye of the Lynx*. Chicago, London: University of Chicago Press, 2002. ISBN 0-226-26147-6 (cloth). The story of Prince Fredrico Cesi, and the formation in early 17th century Italy, of what is arguably the first scientific society. This book is a masterpiece.

Garzanti, (Battaglia, Salvatore, et al.): *Grande dizionario della lingua italiana*. Torin: Garzanti, 1961- <in progress>. Not yet complete, this multi-volume dictionary is the Italian counterpart of the OED. 21 vols published as of 2002. While out of reach most individuals, abridged versions are available on CD, or for free on the website: www.garzanti.it.

Hale, J. R., *Florence and the Medici*. London: Phoenix Press, 1977. ISBN 1-84212-456-0 (paper). A popular account of the Medici dynasty by a great historian.

Hibbert, Christopher, *The Rise and Fall of the House of the Medici*. New York: Morrow Quill, 1974, ISBN 0-688-05339-4 (paper). Solid, detailed popular history of the Medicis.

McLean, Adam, *The Alchemical Mandala: A survey of the Mandala in the Western Esoteric Traditions*, 2nd ed. Grand Rapids: Phanes Press, 1991. ISBN: 1890482951. A great introduction to the symbolism of alchemy by a consumate scholar.

Merrett, Christopher, *The Art of Glass by Antonio Neri*. New York: University Microfilm International, (printed on demand), dist. by www.astrologos.com. This is a direct facsimile print from microfilm of the 1663 translation, owned by Yale University Library.

Moretti, Cesare, ed. *Glossario del Vetro Veneziano Dal Trento al Novecentro* [Glossary of Venetian Glass in the Thirteenth Through Nineteenth Centuries] Venice: Marsilio, 2001. ISBN: 88-317-8030-1. An Italian language glossary of specialized glassmaking terms.

Redi, Fabio et.al, *L'Arte Vetraria a Pisa* [The Glassmaker's Craft at Pisa], Pisa: Pacini Editore, 1994. In Italian with English synopsis booklet. Archaeological accounts of glassmaking sites in Pisa.

Sobel, Dava, *Galileo's Daughter*. New York: Walker and Company, 1999. ISBN 0-8027-1343-2 (cloth) Well written account of early 17th century Florentine life through letters to Galileo from his daughter Maria Celeste.

Zecchin, Luigi, *Vetro e Vetrai di Murano*, (3 Vol). Venice: Arsenale, Vol I: ISBN 88-7743-022-2 (1987 cloth), Vol II: ISBN 88-7743-048-6 (1989 cloth), Vol III: ISBN 88-7743-087-7 (1990 cloth). In his lifetime Zecchin became probably the worlds leading authority on the history of Italian glassmaking. These volumes contain his collected work as published in hundreds of periodicals. His writing exhibits the sharp whit, and skepticism which is so essential for accurate historical research. Sadly out of print, and increasingly hard to find the three volumes together, it is definitely worth the effort. (All in Italian, with a few English Appendices.)

Zupko, Ronald, *Italian Weights & Measures*. Philadelphia: American Philosophical Society, 1981, ISBN 0-87169-145-0. The final word on a very confusing and complex subject.

PICTURE CREDITS

Cover: Jan Wildens (omgeving van), *The Arsenal (Blue Tower) in Antwerp*, Early 17th century. Courtesy of The Netherlands Institute for Art History (Rijksbureau voor Kunsthistorische Documentatie, or RKD)

Back: Blue Tower square, location of Filippo Ghiridolfi's glass shop as it appeared in 1940. Photo courtesy of *Gazet Van Antwerpen* (B.C.) www.gva.be.

The Great Market square (Grote Markt). Attribution unknown.

APPENDIX A

An 1870 Letter by G. F. Rodwell

About an Unpublished Manuscript by Antonio Neri

In the year 1870, the United States was recovering from its devastating Civil War, in Italy Napoleon III had withdrawn French troops from Rome, finally clearing the way for a unified Italian state, and in England, in the September 8th edition of the scientific journal *Nature*, a short correspondence appeared describing an unpublished manuscript (MS.) of the seventeenth century, a manuscript authored by Catholic priest Antonio Neri. Below is the full text of that letter, written by George Rodwell, one time science master at Marlborough College. Neri is best known for his famous 1612 book *L'Arte Vetraria* (The Art of Glass) which was the first ever published on glassmaking technology.

Today, the unpublished manuscript described by Rodwell resides in the Special Collections department of the University of Glasgow Library in Scotland, and is known as *MS. Ferguson 168*. It shows another side of Neri; not concerned with glass, but with more general chemical preparations, and medicinal cures. Happily a date appears on p.265 helping to fix the exact time of its writing: January 26, 1613, a timeframe suggesting these subjects may have been Antonio Neri's central interest in the days preceding his untimely death, about a year later, at the age of 38. Rodwell's letter is an insightful look at Priest Neri, his contemporary political environment, and the nascent state of chemistry in the late Renaissance.

George Farrer Rodwell was born in 1843; he became a fellow of the Chemical Society of London, science master at Marlborough College, and lecturer at Guys Hospital. He traveled extensively through Europe and especially Italy, writing several travel guides, general interest books on chemistry, history, and geology as well as numerous technical papers for the scientific journals. He is an engaging writer, and presents an interesting 19th century perspective of Neri, nicely spanning the worlds of Medician Tuscany and today.

I have endeavored to make Rodwell's letter more easily accessible for the modern English speaking reader by translating the Latin, Italian, and

French passages, which are enclosed in square brackets [ ]. I have also inserted some minor notations, also in square brackets, where it seemed appropriate, with the sole intent of making Rodwell's meaning clear. The original vocabulary and syntax is left intact, as it appeared in 1870. Finally, and perhaps most important, is for the reader to keep in mind that history is a moving target, and that our understanding of Antonio Neri and his life has changed since this letter was first written almost a century and a half ago, and continues to evolve.

Specifically, Neri's birth date is now known to be February 29, 1576, a second unpublished Neri MS. from much earlier in his life has since surfaced (*MS. Ferguson 67*), and letters by his friend Emanuel Ximenes makes it clear that Neri *did* have some kind of responsibilities as a priest within the Church. Also, the biographical dictionary entries mentioned in Rodwell's first paragraph include details that have been hotly disputed by noted glass historian Luigi Zecchin. For instance, the assertions that Neri "traveled all over Europe" and that he deceitfully posed as a "common assistant" in order to learn scientific secrets that he could not gain access to by other means. While the source materials of these original biographers are lost to the mists of time, Zecchin is critically skeptical of these accounts, finding no corroborating evidence whatsoever of Neri's extensive travels, or of his alleged deceptions.

NATURE
Sept. 8, 1870
p. 370-371

ON AN UNPUBLISHED ITALIAN MS. OF THE SEVENTEENTH CENTURY

Not long ago I acquired an MS., entitled *Estratto del Libro, segniato A[B,D,E,F]; di Prete Antonio Neri*, [Excerpts of the books lettered A, B, D, E, F; of Priest Antonio Neri] which I think is of sufficient interest to merit a short notice in these pages. In the first place, let us make ourselves familiar with the life and writings of the priest Antonio Neri. The greater number of biographical dictionaries do not even mention him, but the "Biographie Universelle," and Hoëfer's "Nouvelle Biographie Générale" are exceptions, both giving a short account of him. As to the time of his birth, I can nowhere find a more definite date than *vers le milieu du seizième siècle* [toward the middle of the sixteenth century], while Poggendorff alone mentions the year of his death, which occurred in 1614. [These authors suggest] Antonio Neri was born in

Florence, and was educated for a priest; but he appears never to have undertaken priestly duties, preferring to devote his time to chemical and physical studies. For the purpose of extending his knowledge in this direction, he travelled all over Europe, collecting scientific *secrets*, as they were then called, and he succeeded in amassing a large number of these. He visited the principal laboratories of Europe, and resided for some length of time in Antwerp, where he wrote his treatise on the art of colouring glass. He did not hesitate to work as a common assistant, performing the most menial operations of the laboratory, when he found it impossible to gain access to the secrets he sought by other means.

The period in which Antonio Neri lived coincides most nearly with that of his countryman Baptista Porta [Giambattista della Porta] (born 1537, died 1615). Paracelsus died about the time of the birth of Neri; Jerome Cardan died when he was a boy; Van Helmont was thirty-five years old when Neri wrote his "L'Arte Vetraria [The Art of Glass];" Galileo and Francis Bacon had not reached the summit of their fame; Robert Fludd was busy with his "Historia Macrocosmi [History of the Macrocosm];" Glauber was a boy, Kunckel an infant, Becher was unborn. The Paracelsian iatro-chemistry [medicinal chemistry] was making way,

Crollius was supporting it, while Libavius was the leader of the opposition; the famous "Tyrocinium Chymicum [Chemical Essays]," of Beguinus, was about to appear; the "Academia Secretorum Naturæ [Academy of the Mysteries of Nature]," founded by Baptista Porta, had just been dissolved by Pope Paul V., on the ground that magical and unlawful arts were practiced by its members; but the proceedings of this first of the scientific societies remained in the treatise "Magiæ Naturalis [Natural Magic]," which was the most popular scientific book of the period. Such was the state of the scientific world when Neri laboured and wrote.

The only work ever published by Neri was the treatise on glass-making, to which I have referred. This is a small quarto of 114 pages, and is entitled: "L'Arte Vetraria distinta, in libri sette del R. P. Antonio Neri Fiorentino. In Firenze, 1612. Con licenza di Superiori." [The Art of Glassmaking in seven parts by the Reverend Priest Antonio Neri, Florentine. In Florence, 1612. With Permission of the Inquisition.]

The order of the ecclesiastical authorities for printing this work is conveyed in no less than seven forms, which are signed, countersigned, attested, and endorsed, and bear dates ranging between March 30 and April 7,

1612. This excessive scrutiny may appear strange at first sight, but let us glance for a moment at collateral facts. The [Church's banned book list] "Index Expurgatorius" had been established by [Pope] Paul IV in 1559, and in the very first issue no less than sixty-one printers had been condemned, and the reading of works which issued from their presses forbidden. Now, there was still greater need for caution on the part of the Church; for had not a certain fellow-countryman of Neri, named G[i]ordano Bruno, recently propagated all sorts of heresies in his [book] "Cena de li Ceneri [Ash Wednesday Supper]," and had he not suffered death for his temerity? And was there not a contemporary and countryman of Neri, whilom [former] professor of mathematics [Galileo] in the University of Padua, who had shown particular relish for the doctrines of Copernicus, and a particular disrelish for those of Ptolemy and Aristotle, and altogether an insufficiency of respect for the Church? Thus it was that the utterly inoffensive "L'Arte Vetraria" came to undergo so much scrutiny, and after having been certified to contain nothing *contra fidem aut bonos mores* [against faith or good morals], to be printed *Con licenza di Superiori* [With Permission of the Authorities] on the title-page.

At the same time, I do not at all mean to assert that scrutiny was unnecessary in regard to the scientific works of this period; for although they did not often contain anything *contra fidem*, they very frequently did contain a good deal *contra bonos mores*, in the form of invocations wherewith to raise a familiar demon, recipes for love philtres [potions], and for ingenious draughts for ridding wives of jealous husbands, while the more philosophical "Elixir Vitæ" [health drinks] sometimes required the blood of a new-born babe. I recently met with an alchemical MS., evidently of some rarity, for it was written on vellum, and the binding showed that it had once been in the library of a Doge of Venice: *Recipe Sanguinis Humani* [Human Blood Remedy] were the first words that presented themselves to the eye. Again, [alchemist Johannes] Beguinus says: "*Recipe quantitatem satis magnam sanguinis virorum sanorum, in flore ætatis constitutorum, pone in vase circulatorio, justæ capacitatis, in B. M. continue bulliens, donec draco propriam caudam devoraverit.* [Use a sufficiently large quantity of healthy human blood, taken in the blossom of youth, collect it into a round vessel, of adequate capacity, boil continuously in a Bain Mare, until the serpent eats its own tail[1]]" So that, after all, [novelist] Alexandre Dumas has not given us such a very exaggerated character in the person of the Alchemist Althotas in his "Mémoires d'un Médecin [Memoirs of a Physician]." As to the matter of remedies, I find the following in an MS. in the Sloane Collection

[British Museum] (probably so late as the first half of the 17th century): "Rock crystal, mixed with [the deadly poison] sublimed arsenic, is an excellent medecine; in fact, you need not [take] any other medecine it being taught [to] a witch by a demon, named Rachiel, who was of ye order of Cherubins [fallen angels]." The quaintness and naïveté of this assertion are quite refreshing; now, whether "you need not any other medecine" because the remedy had the sanction of both a witch and a demon, or because powdered rock-crystal and sublimed arsenic had been found by the asserter to be peculiarly adapted for internal administration, we will not pretend to decide; but, surely, so-called scientific books sometimes required examination in the age of the priest Antonio Neri.

Let us, now that we know something of its author, turn our attention to the MS., *Estratto del Libro, segniato A; di Prete Antonio Neri*. There is good reason to believe that the matter of this MS. was extracted by some seventeenth-century chemist from a larger MS. of Neri, of which he speaks in the preface to "L'Arte Vetraria," and which he had intended to publish had his life only been spared.

The text is Italian, but the work cannot be said to be "writ in choice Italian;" it is rugged, and, of necessity, full of technical terms, and it sometimes passes into a curious kind of Latinized Italian. As to the contents, we have extracts from five of Neri's books: from the book segniato A, 155 pages; from B, 78 pages; [C is omitted;] from D, 5 pages; from E, 13 pages; and from F, 6 pages. Between E and F are inserted 26 pages of "Operationi Copiate da un libro antico qui in Pisa [Recipes copied from an old book in Pisa];" also 8 pages about the *Green Lion* [vitriol, or sulfuric acid], and 10 pages of extremely mystical and unintelligible matter, replete with symbols and Arabic words concerning [the transmutation of metals;] a certain *Donum Dei* [Gift of God; a recipe for making the 'philosopher's stone'[2]].

An account of the subject-matter of the extracts from the book designated [by letter] A will, I think, give a fair idea of that of the whole MS. In the first place, we have an account of mercury, to which metal is assigned no less than thirty-five different names, and twenty-two symbols. The Eastern element, then very apparent in chemistry, is noticeable in such names as Chaibach, Azoch, and Baruchet. Various *fisationi* [fixations] of mercury are described, and the formation of some of its compounds. Gold and silver are next discussed; the latter has fifteen names and ten symbols. The fixation and calcination of gold, the calcination of silver, the solution and tincture of silver, and the conversion of silver into gold are then described. Venus (copper) follows, among the fifteen names of which are Tubalchain, Marchaal, and Cobon, but not Cuprum, or Orichalcum, or Æs Cyprium, which is surprising. Then come iron, lead, and tin; then vitriol, which has seven symbols; sal-ammoniac, which has fourteen symbols; sulphur, which has sixteen; arsenic, antimony, sal-alchali [alkali salt], sal-alembrot [mercury salt], sal-tartaro [tartar salt], sal-anticar [borax salt], and cinnabar. The extracts from book A are concluded by accounts of the calcination of various metals, of the philosopher's stone, and of the work of transmutation. The short extracts from the other books contain matter of a similar nature; various well-known salts are described, together with new and varied modes of making them; and different solutions of the metals, compounds, and operations.

The ideas suggested by this MS. are manifold. We can but be struck by the excessive complexity of chemistry at this period. When a substance possesses more than thirty distinct names, and more than twenty symbols, and when these are used indiscriminately in one sentence, some idea may be formed of a chemical treatise of two centuries and a half ago. Symbols were used lavishly, not alone to express substances both simple and compound, but for operations and instruments. But the alchemists and old chemists had a special object in preserving the mysticism out of which their science had sprung, and which still, as a thick vapour, shrouded it in obscurity. Their precious *secrets* would otherwise have been at the command of the vulgar, and the result of their years of toil would have been sown broadcast over the world. The true science was but just beginning to loom through the darksome mists which surrounded it. At this time the science was made up of alchemy and iatro-chemistry, with a strong flavour of Kabbalism [Jewish mysticism]. As to the matter itself, we find in the works of the period scarcely anything more than had been enunciated by [the alchemist] Geber some eight centuries earlier; in fact, there was too much beating about the bush to allow of any real progress. Antonio Neri was a somewhat sensible chemist for the period. His leanings towards alchemy were not excessive; he was not a violent Paracelsian; indeed, he was rather a metallurgical chemist than an iatro-chemist.

Such is a brief sketch of an MS., the matter of which, in a completer form, the priest Antonio Neri had intended to publish had his life been longer spared. Whether the original MS. exists we know not. Perchance it may be hidden in some old monastic library among volumes of Canon Law and countless folios of Middle Age Casuistry [ethics debate]; perchance in some dusty nook in *Ædibus Vaticanis* [the Vatican temple], among the thunderbolts of a past Hierarchy. Who can tell? Oh! if some Sovereign Pontiff would issue a mandate "apud S. Petrum sub annulo Piscatoris [sealed with Saint Peter's ring]," to command the cataloguing of the library of the Vatican, how would not Literature, and Science, and Art be benefited by the means! and how would not Italy receive yet greater honour as the focus from which emanated the glorious light of Western civilization!

GEORGE FARRER RODWELL

---

1. "...until the serpent eats its own tail." a reference to the Ouroboros, one of the oldest mystical symbols in the world. The image of a snake or dragon eating its own tail appears in Egyptian, Greek, Norse, Aztec, Chinese, Occidental and Native American mythologies. The symbol is also appears in the Gnostic, Christian, Hindu, and Ashanti Religions. In Alchemy, the symbol had different meanings in different contexts, often carrying connotations of rebirth, cyclicality, renewal, or the unification of opposites (head and tail). In this context, from a purely technical perspective, the reference may well denote some physically observable change in the liquid preparation as it is heated. Blood is a aqueous colloidal suspension of cells and proteins which at some point will reach a temperature of denaturation, when the constituents start to break down or coagulate; like what happens when an egg is emptied from its shell into hot water. One can speculate that convection currents might form, carrying the suspended solids around in patterns reminiscent of writhing snakes, but this is beside the point. As to the sheer horror of the overall proposition, nothing further need be said, although Rodwell does make an interesting assertion here: while the works of brilliant minds such as Galileo, della Porta, and Bruno were suppressed by the Church, so also were suppressed calls to administer arsenic as medicine, and the wanton bloodletting of innocent children - at a time when such calls just might be taken to heart in certain quarters. Thankfully, Rodwell stops well short of characterizing the Church's Inquisition as some sort of beneficial moderating institution in civil society.

2. "Donum Dei" (Gift of God) is a well-known alchemical work comprised of 12 or sometimes 13 allegorical illustrations. Widely copied and interpreted by numerous authors throughout Europe, its origin is commonly attributed to George Aurach in 1475. Donum Dei draws much of its notoriety as being the recipe of the fabled 'Philosophers Stone' for those adepts able to decipher its mystical symbolism. Descriptions of the 'stone' have been attributed to Aristotle, and the much earlier traditions of Egypt, China, Mesopotamia, and India. It was the 'holy grail' of alchemists throughout history; a material that without itself being consumed, could purify base metals into gold, cure disease, and provide eternal life. Apparently Rodwell was either unaware of the significance of this reference, or was simply content to let it stand without further elaboration.

"Always examine the colors to get to know them by eye, as I have always done, because in this matter, I cannot give specific doses. Sometimes the powder will tint more, other times less, therefore you must practice with your eyes to understand the colors."

- Antonio Neri chap. 95

APPENDIX B

Weights, Measures & Miscellany of 16th - 17th Century Tuscany

Varied Between Towns and Regions:
(Used in Florence, Pisa, Rome, Volterra)
1 Roman pound (libbra; lb) = 12 once = 339 grams
1 Roman ounce (once; oz) = 1/12 libra = 28.25 grams
1 Roman pennyweight (denaro; pwt) = 1/24 once = 1.18 grams
1 Roman grain (grano; gr) = 1/24 denaro = 0.049 grams

In Liguria 1 libbra = 317 grams
In Massa 1 libbra = 332 grams
In Murano 1 libbra = 321 grams
In Piedmont 1 libbra = 307 grams

1 modern (avoirdupois) pound = 16 ounces = 453.6 grams
1 modern (avoirdupois) ounce = 1/16 pound = 28.35 grams
1 modern (avoirdupois) pennyweight = 1/20 ounce = 1.418 grams
1 modern (avoirdupois) grain = 1/7000 pound = 0.0648 grams

Florentine New Years day (secular): March 25th
By Papal decree, the day after Thursday, 4 October, 1582 was Friday, 15 October.
Leap Years: before 1582, all years divisible by 4, but using January 1 as new years day;
After 1582 use the same rules as today.

Dates before 15 October, 1582 add 10 days = Those dates on our current calendar
Dates on our calendar before 15 October, 1582 subtract 10 days = As they would be written then.

Currency:
1 Florin = approx. 28mm dia.

Neri's Hierarchy of Glasses:
Common Glass - (*vetro commune*) - the lowest quality glass.
Crystallino - (*cristallino*) - medium quality, between common and crystal.
Crystal - (*cristallo*) - high quality glass, exceptionally clear and workable.
Artificial Rock Crystal (*bollito, cristallo [di montagna] artificiale*) - finest crystal glass.

"Do not presuppose that I have described a way to make something ordinary, but rather a true treasure of nature, and this for the delight of kind and curious spirits."

- Antonio Neri chap. 133

INDEX

A

absorbed, 49
abundance, 39
access, 6
acetate, 16
acid, 70
    ammoniac, 40, 51
    aqua fortis, 54
    aqua regis, 58, 70
    corrosives, 61, 76
    nitric, 70
    parting water, 70
    vinegar, 5, 7, 8, 9, 16, 17, 28, 29, 51, 53, 56
acquired, 40
adhere, 42
adhesions, 15
advice, 14, 73
affix, 42
agitate, 50, 60, 62
Agostino Vigiani, 77
air, 3, 48, 66
alcohol, 39
alemagna, 35, 39, 75
alembic, 39
almonds, 40
alum, 36 - 38, 40, 45, 48, 49, 50
    saturated, 48
amethyst, 70
ammoniac, 40, 51
amount, 4, 8, 25, 38, 43, 44, 49, 51, 61
ample, 18
amply, 65
ampoule, 40
amusements, 42
ancient, 5
Antonio, 1, 21, 35
Antwerp, 6, 18, 44
anvil, 27
appeal, 1, 15, 25, 27, 30, 32, 56
appear, 4, 42, 70
approaches, 3
approximately, 16, 27, 45, 61, 62
April, 77, 78
aqua, 54, 58, 70
    aqua regis, 58, 70
    aqua fortis, 54
aquamarine, 1, 19, 66, 68, 69
archdiocese, 77
area, 5, 41, 42, 54
around, 41, 60
arsenic, 40
art, 1, 21, 35, 36, 53, 66 - 68, 70, 78
artificial, 4, 69
artisan, 14
artist, 25
avoid, 6
aware, 14
awash, 60
azure, 26, 39, 65, 72, 75

B

backed, 6, 10
bake, 5, 37, 69
balas, 52, 72, 76
ball, 17, 40 - 42, 60, 75
barilla, 67
bars, 59, 60
base, 23 - 33, 42, 54, 57
basic, 4
basin, 43, 44, 46, 47, 48, 50, 53
bathe, 41, 42
bathed, 43
biacca, 15, 16
biadetto, 44
bit, 13, 18, 28, 30, 47
black, 17, 28, 29, 30, 43, 47, 59, 60, 62, 66, 71, 74
blank, 33
blend, 4 -13, 22, 25, 26, 27, 29, 30, 31
blood, 52, 56, 58, 76
blossom, 38, 71, 75
blue, 10, 14, 26, 32, 35 - 42, 44, 45, 61, 62, 64, 65, 68, 69, 74, 75, 76
body, 51
bohemia, 14
boil, 13, 22, 30, 36, 41, 43, 45, 46, 47, 56, 57, 59
bollito, 67, 69
book, 1, 2, 18, 21, 22, 35, 54, 67, 70 -74
borage, 38, 75
bottles, 30, 31
bottom, 5, 16, 17, 22, 37 -40, 44, 46 - 48, 51, 53, 56, 59, 61 - 63, 66
bowl, 41, 43, 53
bran, 46
Brazil, 49, 75
break, 5, 6, 15, 19, 51, 63, 64
bring, 47
broken, 15, 63
bronze, 3, 40, 46, 60, 62, 63
broom, 35, 36, 38, 74
bubbles, 9, 16, 19
bugs, 46
built, 6, 9 -13, 40
bulk, 22
burning, 3, 40, 59, 60
burnish, 40

C

calcine, 3, 16, 17, 18, 19, 22, 32, 42, 51, 54, 55, 56, 58, 59, 60, 61, 62, 63, 68, 70, 71
calx, 22, 23, 54, 55, 57
capable, 15, 64
care, 3 - 5, 15, 28, 60, 65

carnations, 38, 75
carried, 22
carries, 69
carry, 22, 37
case, 3, 49, 60
cast, 19, 40, 41, 43
caution, 67
celestial, 10, 11, 14, 17, 32, 42, 43, 68, 72, 73, 74
ceruse, 15, 16
chalcedony, 2, 57, 68, 70
chance, 15, 48
character, 3
charge, 6, 7, 39, 44, 64, 72
check, 24, 32
chemical, 1, 15, 19
chemist, 15
chimney, 57
christian, 77, 78
chrysolite, 1, 9, 10, 17, 73
churn, 62, 63
cinnabar, 16, 17
clarify, 5, 19, 20, 23 - 32, 47, 52, 54 - 57
clay, 43, 51
clean, 3, 6, 19, 22, 25, 43, 44, 46 - 48, 50, 51, 55, 61, 62
clear, 2, 3, 15, 21, 22, 23, 37, 41, 43, 44, 45, 48, 50, 52, 53, 55, 65, 67
cling, 9, 48
close, 64
closely, 3, 4, 20, 28
cloth, 38, 45 - 48
coagulant, 48
coals, 3, 42, 59, 60, 63
coating, 49, 65
coaxing, 25
cold, 64, 65
collect, 47, 61
color, 1, 3 - 11, 13 - 25, 27 - 33, 35 - 44, 46 - 64, 66, 68 - 76
   alemagna, 35, 39, 75
   amethyst, 70
   aquamarine, 1, 19,

color (cont'd)
   66, 68, 69
   azure, 26, 39, 65, 72, 75
   balas, 52, 72, 76
   biadetto, 44
   black, 28, 29, 30, 43, 47, 59, 60, 62, 66, 71, 74
   blood, 52, 56, 58, 76
   blue, 10, 14, 26, 32, 35 - 42, 44, 45, 61, 62, 64, 65, 68, 69, 74 - 76
   borage, 38, 75
   celestial, 10, 11, 14, 17, 32, 42, 43, 68, 72 - 74
   chalcedony, 2, 57, 68, 70
   chrysolite, 1, 9, 10, 17, 73
   emerald, 1, 4 - 9, 17, 19, 20, 65, 69, 72
   garnet, 1, 12 - 14, 17, 20, 70, 72, 73
   girasol, 72
   golden, 14, 67, 70, 72
   gray, 55
   green, 4, 6, 7, 14, 26, 27, 35, 38, 41, 69, 72, 74
   greenish, 23, 24
   greens, 38, 69, 72, 75
   irises, 37, 38, 74, 75
   jacinth, 1
   lattimo, 71
   lilies, 38
   magpie, 32, 68, 72
   mallow, 38, 75
   milk, 23, 24, 53, 74
   opal, 72
   orange, 38, 75
   peach, 71
   pearl, 71
   pimpernel, 38, 75
   poppies, 37, 38, 74, 75
   purple, 31, 35, 52,

color (cont'd)
   59, 74
   purplish, 31
   red, 14, 16, 17, 30, 31, 35 - 38, 40, 41, 47, 51, 52, 54, 55, 56, 57, 58, 61, 62, 63, 68, 69, 71, 74, 75, 76
   reddish, 52, 60
   roses, 37, 38, 74, 75
   rosichiero, 35, 36, 54, 55, 56, 57, 59, 70, 76
   rubies, 14
   rubino, 58
   ruby, 51, 72
   ruby-red, 51
   sapphire, 1, 11, 12, 14, 17, 19, 71 - 73
   sky, 14, 69
   snow, 17
   tawny, 60, 62, 63
   topaz, 1, 9, 17, 19, 72, 73
   trebiano, 36
   turquoise, 24, 35, 40, 69, 74, 75
   ultramarine, 35, 36, 42, 44, 75
   uncolored, 36, 47, 50
   velvet, 29, 30
   violet, 10 - 13, 33, 37, 38, 73 - 75
   viper, 72
   white, 16, 17, 23 - 28, 30, 31, 36 - 40, 45, 48, 52 - 55, 59, 65, 68, 71, 74, 75
   willow, 47
   yellowish, 16, 18, 52
colorant, 18, 21, 25
common, 17, 19, 21, 22, 35, 37, 41, 43, 50, 51, 52, 56, 60, 61, 63, 68
concentrated, 32, 49
cone, 48
conscience, 77, 78

consolidated, 23
constricted, 48
consume, 6 - 8, 10
contamination, 44
contents, 67
cook, 5, 6 - 9, 11 -13, 15, 25, 26, 31, 52
cool, 45, 47, 59, 61, 63, 65
cool, 2, 16, 43
copper, 3, 14, 19, 24, 26, 27, 28, 31, 33, 40, 52, 54, 55, 56, 57, 59, 60, 61, 62, 63, 65, 69, 76
copy, 77
Corn, 78
corrosives, 61, 76
cost, 69
could, 65, 66
counter, 77
cover, 3, 5, 9, 16, 23, 39, 40, 46, 53, 59, 70
cracked, 63
crocus, 4, 7, 8, 9, 27 - 29, 54, 57, 58, 68
crucible, 3, 5 - 13, 15, 19, 23 - 33, 52, 54 - 57, 59
crumble, 3
crush, 3, 4, 46
crystal, 2 - 4, 6 - 13, 18, 19, 23, 39, 51, 52, 54, 56 - 58, 66 - 69, 71, 72
crystalline, 65
crystallino, 68
crystals, 65
cull, 42
curiosity, 78
curious, 14, 25, 36, 66
customs, 77, 78

D

damp, 37, 65
dark, 59
darken, 44, 52
day(s), 15, 17, 19, 38, 40, 41, 43, 45, 50, 54, 58, 61 - 63, 66, 75

decant, 16, 22, 37, 43, 44, 46, 47, 53, 61, 62, 63, 65
decompose, 3
definite, 59
delight, 66
demonstration, 21, 51, 54
deposits, 51
depth, 5, 12, 14, 16, 53, 73
described, 2, 4 - 7, 9, 11, 13 -15, 18, 19, 21, 22, 24, 26 - 29, 31, 32, 38, 43 - 46, 54 - 57, 60, 62, 64, 66
description, 50
desired, 1, 3, 25, 38, 49, 54, 55, 58, 75
desks, 42
detached, 63
diaphanous, 51
die, 50, 65
digest, 45, 54
diligence, 2, 4, 5, 9, 16, 20, 21, 60
dirt, 15
discretion, 14, 30, 31, 52, 55
dissolve, 17, 36, 43, 48, 50, 70
distill(ed), 16, 17, 39, 51, 53
doctrine, 2
dollops, 24, 27, 29
dose(s), 6, 7, 8, 18, 24, 27, 28, 37, 45, 57, 66
drain, 37, 41, 47, 48, 56
dram, 55
draws, 36
dregs, 17, 20, 52, 70
dress, 3
dried, 3, 16, 46, 62
dust, 23, 47
dye, 37, 38, 39, 46 - 50, 63, 64
dying, 50, 65

E

earrings, 14
earth, 3, 63

earthen, 17, 19, 36, 43, 53, 54, 58, 61, 63, 64
earthenware, 22
elixir, 46, 50, 75
embers, 3
emerald(s), 1, 4 - 9, 17, 19, 20, 65, 69, 72
empty, 9, 22, 36, 42, 43, 50
enamel, 10, 21 - 33, 35, 36, 41, 45, 53, 54, 57, 58, 59, 73, 74, 76
end, 14, 22, 42, 45, 47, 51, 57, 59, 60, 62, 63, 65, 66, 76
erupt, 29
essence, 3
evaporate, 16, 17, 22, 53, 61 - 65
evolve, 65
examine, 24
example, 18, 19, 22, 23, 26 - 28, 31, 56, 57, 61, 64
expense, 2, 49
experiment, 58
expertise, 4
extract, 14, 16 - 18, 35, 37 - 39, 46, 47, 49, 53, 61, 63, 64, 67, 74 - 76
eye, 24, 54, 61, 62

F

facet, 9
faded, 35, 40, 75
fenugreek, 46
fern, 67
ferretto, 27, 29, 68
fibers, 48
field, 21
Filippo, 77
fill, 5, 6, 64
filter, 16, 17, 61 - 64
filth, 44, 64
fine, 19, 20, 22, 23, 36, 40, 44, 45, 54, 55, 57, 58, 69
finely, 4, 15, 17, 43, 52, 54
finest, 3, 22, 53
finger, 4, 5, 7, 9, 13, 53
fire, 5 - 10, 12, 13, 16, 17,

fire (cont'd)
    19, 21 - 23, 36, 39, 40, 41, 43, 45 - 47, 51, 54, 56, 59, 60 - 65
fired, 3, 6, 48
firewood, 6, 59
fish, 41
fixing, 7, 55, 56, 59, 76
flake(s), 26 - 28, 31
flaming, 65
Flanders, 14, 15
flask, 16, 17, 45, 46, 50, 64
flat, 54, 59
flint, 2
float, 56
Florence, 77, 78
flour, 3, 4, 17, 43, 54
flower petals, 38, 39
flowers, 35, 36, 38 - 40, 56, 74, 75
    blue irises, 38, 75
    borage flowers, 38, 75
    broom flower, 35, 36, 38, 74
    carnations, 38, 75
    day lilies, 38, 75
    irises, 37, 38, 74, 75
    lilies, 38, 75
    orange blossoms, 38, 75
    ordinary violets, 38, 75
    peach blossom, 71
    poppies, 37, 38, 74, 75
    red poppies, 38, 75
    red roses, 38, 75
    red violets, 38, 75
    roses, 37, 38, 74, 75
    violet roses, 37, 74
    violets, 38, 75
flowers of sulfur, 40, 56
foam, 59
foil, 6, 7, 10, 68
form, 5, 15, 21, 28 - 32, 40, 43, 52, 54, 57, 65
foulness, 17
fragments, 42, 50
frankincense, 43

free, 2, 48
fresh, 3, 36 - 38, 42 - 48, 53
frit, 23, 52, 54 - 57, 67, 71
froth, 9
fumes, 62, 63
furnace, 5, 6, 9 - 13, 15 - 17, 19, 23 - 27, 29 - 31, 33, 54, 58 - 64
fuse(d), 6, 23 - 26, 28, 30, 40, 51, 54, 55, 58

G

garnet(s), 1, 12 - 14, 17, 20, 70, 72, 73
gems, 1, 3, 4, 6, 14, 15, 19, 20, 73
    amethyst, 70
    aquamarine, 1, 19, 66, 68, 69
    balas, 52, 72, 76
    chrysolite, 1, 9, 10, 17, 73
    emerald, 1, 4 - 9, 17, 19, 20, 65, 69, 72
    facet, 9
    garnet, 1, 12 - 14, 17, 20, 70, 72, 73
    girasol, 72
    imitation, 1, 15, 20, 41, 73, 75
    jacinth, 1
    jewels, 2, 5, 18
    natural stones, 1, 2, 20, 41, 75
    opal, 72
    oriental, 1, 3 - 5, 7, 9 - 13, 65, 70, 72, 73
    paste, 1, 2, 4 - 7, 9 - 15, 17 - 20, 22, 42, 43, 51, 54, 72, 73
    pearl, 71
    rubies, 14
    ruby, 51, 72
    sapphire, 1, 11, 12, 14, 17, 19, 71,

gems (cont'd)
    72, 73
    stones, 1 - 3, 14, 20, 70
    topaz, 1, 9, 17, 19, 72, 73
    turquoise, 24, 35, 40, 69, 74, 75
German, 3
girasol, 72
glass, 1, 3, 6, 15 - 21, 23 - 31, 35 - 37, 39 - 42, 45, 48, 50 - 58, 61 - 72, 75, 76, 78
glassmaking, 21, 25, 36, 53, 66 - 68, 70, 78
glass-pots, 6
glazed, 17, 19, 23 - 31, 36, 43, 50, 52, 53, 54, 55, 57, 61, 62, 64
glue, 41, 42
gold, 6, 7, 14, 21, 24, 32, 35, 36, 54, 58, 59, 70, 76
golden, 14, 67, 70, 72
goldsmith, 21, 24, 26 - 33, 59
grain, 3, 4, 7 - 13, 17, 43, 46
grappa, 39, 50
gray, 55
green, 4, 6, 7, 14, 26, 27, 35, 38, 41, 69, 72, 74
greenish, 23, 24
greens, 38, 69, 72, 75
    mallow, 38, 75
    pimpernel, 38, 75
grind, 3, 4, 13, 15, 16, 19, 29, 30, 31, 40, 43, 44 - 46, 51, 57, 59, 60, 62, 63
grinding (see ground), 4
grossly, 48
ground, 4, 6 - 12, 18, 23, 27, 41 - 43, 46, 53, 54, 56, 62, 63
gruma, 30, 31
gum, 43, 44, 45

H

hammerings, 28, 33
hand, 37, 77, 78
hard, 2, 6, 8, 16, 18 - 20, 42, 51, 54, 73
heat, 5, 16, 40, 43, 44, 51, 55, 57, 60
hematite, 55
hole, 19, 42
holy, 77
hook, 60, 63
hot, 16, 42, 43, 44, 60
hour(s), 6, 16, 23, 26, 44 - 47, 51, 54 - 59, 61, 62, 65
hydrated, 41

I

ignite, 60
illuminate, 14
imitate, 1, 15, 20, 41, 73, 75
impalpable powder, 4, 6, 8 - 13, 17, 43, 51
imperfections, 4, 6, 7, 9, 10, 17, 19
imprint, 78
impurities, 15
incombustible, 59
incorporate(d), 4, 9, 24, 26 - 29, 43, 45, 52, 55 - 57
inexpensive, 14
inflamed, 3, 42
infused, 41
ingredients, 46
inquisitor, 78
invention, 2, 14, 50, 67, 69
inverted, 48
irises, 37, 38, 74, 75
iron, 3, 14, 27, 54, 59, 60, 62, 63, 68
Isaac, 1, 2, 15
isinglass, 41
Italian, 3

J

jacinth, 1
jewels (see gems), 2, 5, 18

joints (see lute), 39, 65

K

kermes, 35, 45 - 51, 75
kettle, 22, 26, 28, 31, 33, 43, 45 - 48
kettle-smith, 26, 28, 31, 33
kiln, 6, 9, - 13, 15, 22

L

ladle, 64
lake, 35 - 39, 41, 45, 48 - 51, 74, 75
lapis lazuli, 35, 42 - 45, 72
lattimo, 71
lead, 7, 8, 10, 14 - 17, 22, 41, 52, 54 - 57, 71, 72
leaf, 38
Levantine, 18, 46
llid, 5, 7, 9
lilies, 38
lime, 36, 38, 59
limekiln, 54
linen, 37, 38, 48, 49, 51
linseed, 43
liqueur, 64, 65
liquid, 16, 48, 63 - 65
litharge, 16, 53
loading, 6, 8, 11, 15, 24, 73
lukewarm, 40, 41, 44, 46
lute, 5, 7 - 9, 11 - 13, 16, 39, 40, 59, 65
lye, 36 - 38, 46 - 48, 50, 59

M

madder, 35, 49, 75
magpie, 32, 68, 72
mallow, 38, 75
manganese, 11 - 14, 19, 20, 23, 25, 28, 30, 31, 33, 51, 52, 54, 55, 68
marble, 71

march, 77
martis, 4, 7 - 9, 27, 28, 29, 57, 58, 68
mass, 5, 43
mastic, 43
material, 3 -14, 16 - 19, 21 -33, 38, 40, 42, 43, 49, 50, 53, 54, 57, 58, 73
measuring, 7, 25
medicine, 51
melt, 16, 23, 24, 28, 30, 32, 40, 43, 52, 56, 57, 72
mercury, 40, 70
metal(s), 21, 22, 24, 30, 32, 35, 40, 70
  arsenic, 40
  bronze, 3, 40, 46, 60, 62, 63
  copper, 3, 14, 19, 24, 26 - 28, 31, 33, 40, 52, 54 - 57, 59 - 63, 65, 69, 76
  gold, 6, 7, 14, 21, 24, 32, 35, 36, 54, 58, 59, 70, 76
  iron, 3, 14, 27, 54, 59, 60, 62, 63, 68
  lead, 7, 8, 10, 14 -17, 22, 41, 52, 54 - 57, 71, 72
  manganese (oxide), 11 - 14, 19, 20, 23, 25, 28, 30, 31, 33, 51, 52, 54, 55, 68
  mercury, 40, 70
  silver, 14, 70
  tin, 22, 40, 54, 55, 57
metallic, 22
Migliore, 77
mildew, 48
milk, 23, 24, 53, 74
minerals (see gems)
  alum, 36 - 38, 40, 45, 48 - 50
  cinnabar, 16, 17
  flint, 2
  lapis lazuli, 35, 42, 43, 44, 45, 72
  manganese (oxide),

Minerals (cont'd),
11 - 14, 19, 20,
23, 25, 28, 30,
31, 33, 51, 52,
54, 55, 68
   marble, 71
   minium, 4, 6 - 13,
16, 18, 41, 55
   natural stones, 1, 2,
20, 41, 75
   opal, 72
   porphyry, 3, 4, 19,
40, 43, 44, 45,
51
   quartzite, 2
   sal ammoniac, 40,
51
   salt, 16 - 19, 23, 37,
51, 54, 57, 67,
68
   saltpeter, 40, 51
   stone, 4, 19, 40, 42 -
45, 51
   sulfur, 17 - 19, 40,
55 - 57, 59, 60,
62, 63, 76
   zaffer, 10 - 14, 19,
20, 24 - 26, 28,
29, 32, 66, 68
minium, 4, 6 - 13, 16,
18, 41, 55
mirror, 40, 75
Miserere Psalm, 46
mix, 5, 7 - 9, 11 - 13, 19,
23 - 33, 43, 46, 51,
54, 57, 60, 62, 63,
64, 66
mixture, 25, 29, 40, 75
moisture, 40, 42, 48,
49, 61 - 63
mordant, 45, 46
mortar, 3, 18, 46, 60,
62, 63
mouth, 16, 39, 40, 48
muller, 3
musty, 47

N

natural, 1 - 4, 6, 19, 20,
40 - 42, 70, 75, 78
natural stones, 1, 2, 20,
41, 75

neck, 16, 40, 50, 51
Neri, 1, 21, 35
Nic(c)olini, 77
nitric, 70
noble, 14, 21, 36, 51,
53, 64

O

oak, 6, 23, 59
odor, 66
oil, 40, 43, 56
opal, 72
opaque, 5, 57, 71
orange, 38, 75
oriental, 1, 3 - 5, 7, 9 -
13, 65, 70 - 73
orpiment, 41, 68
ounce, 4, 7, 8, 10 - 13,
49, 51
over-colored, 55
overflow, 31
oxides
   crocus martis, 4, 7 -
- 9, 27 - 29, 54,
57, 58, 68
   iron, 3, 14, 27, 54,
68
   lead calx, 22, 54, 55,
57, 58, 71
   lime, 36, 38, 59
   litharge, 16, 53
   minium, 4, 6 - 13,
16, 18, 41, 55
   Piedmont manga-
nese, 11 - 14, 19,
20, 23, 25, 28,
30, 31, 33, 51,
52, 55, 68
   quartzite, 2
   red copper, 52, 54 -
57, 69
   red lead, 16, 41
   spanish feretto, 27,
29, 67, 68
   tarso, 23, 54
   thrice cooked cop-
per, 19, 24, 26,
27, 28, 31, 33
   tin calx, 22, 54, 55,
57
   zaffer, 10 - 14, 19,
20, 24 - 26, 28,

oxides (cont'd)
29, 32, 66, 68

P

pack, 16, 39, 60
painted, 42
painters, 10, 35 - 39, 45,
49, 75
pale, 40
palm, 16, 69
pan, 3, 17, 19, 43, 44,
51, 54, 58, 61 - 65
paper, 39
Paris, 42
parting water, 70
paste, 1, 2, 4 - 7, 9 - 15,
17 - 20, 22, 42, 43,
51, 54, 72, 73
peach, 71
pearl, 71
pendants, 14
pennyweight, 4
pestle, 4, 46
petals, 38, 39
piedmont, 11 - 13, 23,
25, 28, 30, 31, 33,
51, 52
piero, 77
pigment, 38, 48
pimpernel, 38, 75
pine, 43
Pisa, 25, 49 - 51, 58
pitch, 43
plant(s), 35, 38, 39, 67,
74, 75
plaster, 42
plates, 5
polish, 1, 2, 4, 6, 20, 40
polverino, 18, 54, 67
poppies, 37, 38, 74, 75
porphyry, 3, 4, 19, 40,
43 - 45, 51
pot, 5, 23, 36, 42
potash, 52
potent, 65, 70
pottery, 5, 6, 9, 10, 12,
13, 19
powder(s), 3 - 5, 8, 11,
12, 19, 24 - 33, 40,
42, 43, 55, 57, 58,
60, 62, 66, 69

precipitate, 53
prices, 42
priest, 1, 21, 35
proof, 23, 24, 56, 57
proportion, 7, 10, 11, 12, 13, 43, 49, 58
pulverize, 8, 23, 45, 48, 50, 52, 60, 63, 66
purify, 5, 6, 15, 17, 18, 29, 40, 54, 55, 68, 70
purple, 31, 35, 52, 59, 74

Q

quartzite, 2

R

rags, 48
raw, 45, 46, 55
receiver, 39, 65
recipe, 7, 38, 54, 55, 57
re-cook, 15
recover, 6, 53
red, 14, 16, 17, 30, 31, 35 - 38, 40, 41, 47, 51, 52, 54 - 58, 60 - 63, 68, 69, 71, 74 -76
reduce, 3, 24, 25
reed, 41
re-fire, 5
regent, 77, 78
regis, 58, 70
registry, 78
regrind, 4
reheat, 19, 25, 27, 32, 33, 52, 54, 55, 57, 58
re-lute, 5, 15
remains, 19, 30, 51, 61
residue, 16, 64, 66
retort, 51, 65, 66
reverberating, 51, 57
rings, 14
rocchetta, 67
roche, 36, 45, 48 - 50
rock, 2 -13, 18, 67, 71, 72
rod, 43, 46, 62
root, 35, 49, 75

roses, 37, 38, 74, 75
rosichiero, 35, 36, 54 - 57, 59, 70, 76
round, 43, 59
rub, 38
rubies, 14
rubino, 58
ruby, 51, 72
rumbling, 16

S

sal, 40, 51
salt, 16 -19, 23, 37, 38, 51, 54, 57, 67, 68
saltpeter, 40, 51
sand, 16, 17, 40, 51, 61, 62, 64
sapphire, 1, 11, 12, 14, 17, 19, 71, 72, 73
saturated, 14, 16, 17, 64
Saturn, 17 - 19, 53, 76
scale, 69
scoop, 48, 51
scrape, 48, 60, 62
sear, 70
secret, 67, 70, 78
sediment, 17, 22, 47
servants, 77, 78
settle, 17, 37, 38, 61 - 63
sew, 48
shape, 41, 48
sharp, 65
shearings, 45 - 50, 75
sheets, 27, 78
shelves, 42
shine, 2, 20
sieve, 19, 22, 23, 43, 44, 46, 59, 60, 62
sift, 3, 4, 17, 23, 43, 46, 50, 59, 52, 60, 62, 63
sign, 5, 27, 37, 47
silver, 14, 70
simmer, 56, 57
singed, 40, 57, 58
sky, 14, 69
smoke, 60
smokeless, 23
snout, 39
snow, 17
soak, 43, 46, 51
soda, 36, 38, 67

softwood, 47
solution, 17, 59
soot, 57, 58
space, 5, 7, 8, 9, 13, 30
Spanish, 27, 29, 67, 68
spatula, 60
specs, 6
sphere, 40, 75
spirit, 14, 36, 53, 65, 66, 76
spoon, 4, 41
spread, 37, 47, 48
sprinkle, 41, 58
staining, 39
steam, 65
stick, 47, 60, 62, 65
sticks, 48, 63
stir, 4, 24, 43, 44, 46, 47, 52, 55, 57, 62
stocking, 48, 50
stone, 3, 4, 19, 40, 42 - 45, 51
stones, 1 - 3, 14, 20, 70
store, 23
strain, 47, 48, 50
stretch, 48
strike, 52, 57, 58
sublimation, 40, 51
sugar, 3, 17
sulfur, 17 -19, 40, 55 - 57, 59, 60, 62, 63, 76
sun, 19
suspended, 48, 50
sweat, 18
sweet, 16, 17, 37, 40
swells, 9, 13, 16, 28, 30, 31, 57
swirl, 41
sword, 55

T

table, 67
tar, 43
tarso, 23, 54
tartar, 23, 30, 40, 45, 46, 54 - 58, 68, 70
tawny, 60, 62, 63
tegame, 59, 60 - 63
terracotta, 5, 9, 23, 59
test, 25
thrice cooked, 19, 24, 26 - 28, 31, 33

throw, 3, 23 - 33, 40 - 43, 46, 52 - 58, 64
tiles, 37, 48, 49, 51
tin, 22, 40, 54, 55, 57
tincture, 14, 19, 36, 38, 39, 64, 65
tinsel, 32, 68
tint, 11, 13, 19, 24, 25, 41, 42, 45 - 47, 49 - 53, 56, 61 - 64, 72, 75
tool, 60
topaz, 1, 9, 17, 19, 72, 73
toy, 42
transparent, 3, 5, 35, 36, 51, 58, 68, 75, 76
trap, 48
trays, 17, 27
trebiano, 36
tree, 35
turpentine, 43
turquoise, 24, 35, 40, 69, 74, 75
Tuscany, 59
twigs, 47

U

ultramarine, 35, 36, 42, 44, 75

uncolored, 36, 47, 50
unify, 4, 12, 19, 25, 26, 28, 31 - 33, 70
universal, 21
un-lute, 65
urinal, 15 -18, 39, 61 - 64

V

vapors, 65
velvet, 29, 30
Venice, 42
Venus, 59, 61, 76
verdigris, 4, 7, 8, 14, 41
verzino, 35
vessel, 5, 16, 22, 37, 39, 41, 43, 50, 52, 59, 61, 63, 65, 75
vial, 57, 58
Vicker, 77
vine, 47
vinegar, 5, 7 - 9, 16, 17, 28, 29, 51, 53, 56
violet, 10 - 13, 33, 37, 73, 74
violets, 38, 75
viper, 72
vitrify, 2, 5, 57
vitriol, 59 - 61, 64, 65, 70, 76

W

warm, 43, 44, 47, 51, 60, 62 - 64
warnings, 15, 69
wash, 37, 38, 44, 45, 48, 51
water, 3, 17, 19, 22 - 33, 37, 38, 41 - 58, 61 - 65, 76
waves, 41, 42
wax, 43
weight, 25, 40, 51, 60
white, 16, 17, 23 - 28, 30, 31, 36 - 40, 45, 48, 52 - 55, 59, 65, 68, 71, 74, 75
willow, 47
wind-furnace, 59
wine, 20, 30, 31, 36, 40, 52, 54 - 56, 70, 74
wood, 6, 23, 49, 75
wooden, 42, 48, 51
wool, 45, 50, 75

XYZ

yellowish, 16, 18, 52
zaffer, 10 - 14, 19, 20, 24 - 26, 28, 29, 32, 66, 68